The Prepper's Survival Bible [25 Books in 1]

A Long-Term Survival Guide, from Fundamental Lifesaving Skills to Advanced Proficiency, with Off-Grid Tactics, Stockpiling Secrets, Water Purification, and Defensive Strategies for Any Scenario

By

HUGHIE TIMOTHY MARROW

© **Copyright 2024**

All rights reserved.

This document's objective is to provide accurate and dependable information about the subject and issue at hand. The publisher sells the book with the knowledge that it is not bound to provide accounting, legally permissible, or otherwise qualifying services. If legal or professional assistance is required, it is prudent to consult a knowledgeable specialist.

From a Declaration of Principles that an American Bar Association Committee and a Publishers and Associations Committee jointly recognized and approved.

No portion of this publication, whether electronic or printed, may be reproduced, duplicated, or transmitted in any form or by any means. It is highly illegal to record this publication, and storage of this document is permitted only with the publisher's prior consent.

The material presented below is declared to be accurate and consistent, with the caveat that any liability resulting from the use or abuse of any policies, processes, or directions contained herein, whether due to inattention or otherwise, is entirely the recipient reader's duty. The publisher shall not be accountable for any compensation, damages, or monetary loss experienced as a result of the information included herein, whether directly or indirectly.

The information presented on this page is primarily educational in nature and is hence universal. The information is provided "as is" with no implied commitment or guarantee.

The trademarks are utilized without the permission or backing of the trademark owner, and the trademark is published without the consent or backing of the trademark owner. All trademarks and registered trademarks referenced in this book are the property of their respective owners and are not associated with this publication.

TABLE OF CONTENTS

BOOK 1: INTRODUCTION TO PREPAREDNESS ... 13
- What is disaster preparedness? ... 13
- What is deemed as an emergency or disaster? ... 14
- Ways to prepare for a disaster ... 14
  - More ways to prepare ... 14
- What is the survival/preparedness mindset? ... 15
  - How can I prepare my mind to deal with crises and natural disasters? ... 15

BOOK 2: PREPPING FUNDAMENTALS ... 19
- How to Start Prepping for Beginners ... 19
  - Food Prepping ... 20
  - Water Prepping ... 21
  - First Aid Prepping ... 22
  - Prepping Hygiene and Sanitation ... 23
  - Prepping Survival ... 23
  - How to Prep for Emergencies that Happen Away From Home ... 24
- Learn Core Skills, Practice With Your Gear & Maintain Your Equipment To Prep ... 25
  - Learn Core Skills ... 25
  - Practice With Your Gear ... 26
  - Practice making a fire utilizing different techniques ... 26
- Connect and Share Your Prepping Experience With Other Preppers ... 26

BOOK 3: VARIOUS SCENARIOS TO PREPARE FOR ... 28
- Natural Disasters ... 28
- Human-Caused Disasters ... 29
- Prepping for Wildfires: A Beginner's Guide ... 29
- Prepping for Hurricanes ... 31
- Prepping for Tornadoes: A Beginner's Guide ... 31

    Blizzard Prepping for Beginners: How to Prep for the Cold 32

    The Essential Beginner Prepper Checklist 33

BOOK 4: CRISIS PREPAREDNESS GUIDE 35

    The four stages of a crisis 35

        Pre-crisis phase 36

        Crisis phase 36

        Response phase 37

        Post-crisis phase 37

    Building a comprehensive crisis management plan 38

        Proactive 38

        Responsive 38

        Recovery 39

        Emergency Evacuation Plan 39

BOOK 5: MENTAL HEALTH SURVIVAL GUIDE 42

    Understand Disaster Events 42

    Recognize Signs of Disaster-Related Stress 43

    Easing Stress 43

    Helping Kids Cope with Disaster 44

        Recognize Risk Factors 44

        Vulnerabilities in Children 44

        Meeting the Child's Emotional Needs 45

        Reassuring Children After a Disaster 46

        Monitor and Limit Exposure to the Media 46

        Use Support Networks 47

    Preparing Emotionally for a Disaster or Emergency 48

        How can I build emotional wellness? 48

        How can I build a strong support network? 49

        How can I take better care of myself? 50

Taking care of practical details ahead of time can help lower stress during an emergency. 50

Helping Your Child or Teen Prepare Emotionally for a Disaster or Emergency 51

BOOK 6: FOOD PREPAREDNESS AND PREPPER'S PANTRY 54

What Kind of Food Should You Store in a Prepper Pantry? 54

Food you can buy in bulk 55

Food that's easy to prepare 56

Preserving Your Own Food 56

How About Freeze Dried Food? 57

Selecting the Right Place 57

Building a Prepper Pantry With Limited Space 58

What Kind of Storage Supplies Do You Need? 58

Mylar bags 58

Food buckets 59

Oxygen absorbers 60

How to organize your prepper pantry 61

FIFO Method 61

Using Can Dispensers 61

Disaster-Proofing Your Shelves 61

Have a Working System 62

Keep an Inventory of Your Supplies 62

BOOK 7: PREPPER'S COOKBOOK 64

Basic Pioneer Cornbread 64

Hardtack 65

Pemmican 66

Native American Fry Bread 67

Biltong 68

Cornmeal Mush 69

Bannock 70

Red Beans and Rice	70
Pinole	71
Dumplings	72
Corn Dodgers	73
Pan Fried Pork Chops	74
Potato Cakes	75
Oxtail Potjie	76
Fish on a Stick	77
Vetkoek	78
Rusks	79
Buffalo Jerky	80
Pioneer Bread	81
Johnny Cakes	82
Venison Stew	83
Dried Fruit	84
Peasant Bread	85
Hasty Pudding	86
Old Fashioned Stack Cake	87
Emergency Cooking	88
Knowledge	89
Skills	89
Equipment	89
BOOK 8: FIRE-MAKING MASTERY	**93**
Fundamentals of Fire Building	93
Choosing a Site	93
Fire Material Selection	93
Arranging Your Materials	94
Fire Building Techniques	94
Ignition Methods	95

    Fire Maintenance 96

BOOK 9: PREPPER'S NATURAL MEDICINE 99

Types of Herbs to Utilize 99

    Upset Stomach 99

    Congestion 100

    Burns 101

Three types of Herbs to Have on Hand for Emergencies 102

BOOK 10: PREPPER'S WATER SURVIVAL GUIDE 104

Stocking Water 104

Indoor Water Sources 105

    What if the water coming into the house is contaminated? 106

Outdoor Water Sources 107

Water Purification Treatment 108

Book 11: Off-Grid Living Essentials 110

off-grid solar energy 110

    Benefits of Solar Power Systems for Off-Grid Living 110

    Components 110

    Solar system design 115

    Determine your energy needs 117

    Selecting Batteries 120

    Solar panel array sizing 122

    Inverter sizing 123

Wind Power 123

    Why Use Wind Power? 124

    How to Use Wind Power 126

Hydropower 127

    What is a Micro Hydropower System? 127

    How to Measure Hydrosystems Head? 129

How to Measure Hydrosystems Flow? 130

Self-Sufficient Living Skills and Practices 132

Book 12: Prepper's Home Defense 136

Why Is It Necessary to Fortify Your Home? 136

Then, how precisely do you defend your house? 136

BOOK 13: RV CAMPING FOR SURVIVAL 143

Mobility and Rapid Response: 143

Emergency Shelter and Accommodation: 143

Resource Delivery and Distribution: 143

Communication and Connectivity: 144

Community Support and Rehabilitation: 144

Before Disaster Strikes 144

    Fill Your Tanks 145

    Know your surroundings 145

    Have a Reliable Alert Source 145

    Stock up on Essentials 146

General RV Care Tips 146

BOOK 14: CREATING COMMUNITY AND SHARING YOUR PREPPING 150

The Benefits of Prepper Communities or Building a Survival Network 150

Skill Diversification: Building a Network to Survive! 151

A Good Survival Network Can Provide Emotional Support 152

Finding or Forming a Prepper Community 152

BOOK 15: ENERGY INDEPENDENCE FOR SURVIVAL 155

The Role of Renewable Energy in Disaster Preparedness 155

    Reliable and Resilient Power 155

    Self-Sufficiency and Energy Independence 155

    Off-Grid Power Generation 156

The benefits of including renewable energy in plans for disaster preparedness 156

Strategies for Sustaining Energy Efficiency 157

DIY Energy Projects for Off-Grid Living 158

    Project (building a solar off-grid power system) 158

    Project (building a micro-wind turbine system) 164

BOOK 16: WILDERNESS SURVIVAL TECHNIQUES 166

Build a fire 166

Build a shelter 166

    Identifying suitable shelter locations 166

    building several kinds of shelters according to the resources at hand. 166

    insulating the shelter to keep the weather out and provide warmth. 167

    Build an Off Grid Cabin 167

Create a hierarchy of importance. 169

Locate a source of pure water. 169

Safeguard food and wild plants 169

    Identify edible plants, berries and insects in the wild 169

    Learn the fundamentals of fishing and trapping. 169

    Recognizing the significance of wilderness food safety 170

Maintain impeccable hygiene. 170

Remain composed and weigh the circumstances. 170

Alert search and rescue personnel in the area. 171

Examine your bushcraft abilities in advance of using them. 171

Make use of everything available to you. 171

BOOK 17: COMMUNICATION IN CRISIS 173

Communication Options in a Disaster 173

    Landline Phones 173

    Two-way radios 174

BOOK 18: PREPPER'S FIRST AID MANUAL 176

First Aid Kits 176

    Natural Disaster. 176

    Medical Emergencies. 177

  Medical Supplies and Equipment for First Aid Kits 177

    Put an end to a bloody nose. 178

    Treat a Cut Finger 178

    Handle a Sprain 178

    Take out a Splinter 178

    Put an end to diarrhea 178

    Address nausea 179

    Head Lice Killing 179

    Handle insect bites 179

    Handle a Burn 179

  ABCs of First Aid 179

  First Aid Kit List 180

Book 19: URBAN SURVIVAL STRATEGIES 182

BOOK 20: SHELTER AND SANITATION SOLUTIONS 185

    Shelter 185

  Hygiene 188

    Effective Hand Washing 188

  Sanitation 189

    Potty Options 191

Book 21: Prepper's Gardening Guide 193

  The Self-Sustaining Food Garden 194

    Planning a Self-Sufficient Garden 195

    Setting mental expectations 195

    Some tips 196

BOOK 22: BONUS CONTENT 1: ADVANCED SURVIVAL TECHNIQUES 200

how to make and spear fish 200

How to Find Water in the Wild 201

    Start With the Obvious: Streams, Rivers, Lakes 201

    Collect Rainwater 202

    Collect Heavy Morning Dew 203

    Fruits/Vegetation 203

    Collect Plant Transpiration 203

    Tree Crotches/Rock Crevices 203

    Dig an Underground Still 204

    Melt Snow and Ice 205

    Avoid Water Substitutes 205

Practice knife throwing 206

BOOK 23: BONUS CONTENT 2: ADVANCED SURVIVAL TECHNIQUES 2.0 210

How to Read a Topographic Map 210

    Topographic Map Lines, Colors and Symbols 211

    Topographic Map Contour Lines 212

    Topographic Map Scale 213

how to use a compass 214

    Essential Parts of the Compass 214

    Setting Declination 215

    Taking a Bearing from a Map 215

    Taking a Bearing in the Field 216

    Triangulation 217

How to Raise Honeybees in Your Backyard 217

    Pest control 220

    Control methods 220

    Honey collection 222

BOOK 24: BONUS CONTENT 3: FAMILY PREPAREDNESS 223

- Identify hazards 223
- Hold a family meeting 223
- Prepare 224
- Practice your plan 224
- How to Guide Your Children During a Disaster 225
- How to Help Your Child After a Disaster 225
  - How to Proceed. 225

BOOK 25: BONUS CONTENT 4: PREPPER'S RESOURCE DIRECTORY 227
- Explore the Top Survival Forums and directories to Follow 227

CONCLUSION 230

BONUS 231

BOOK 1: INTRODUCTION TO PREPAREDNESS

Disasters occur on varying scales throughout the year, in every part of the world, and at all seasons. During these catastrophes, nonprofit groups play a crucial role in helping the affected population by offering resources and assistance. In fact, during most emergency situations, hospitals, food banks, shelters for people and animals, and other charitable organizations are included in the response and recovery processes.

By their very nature, these events are frequently unanticipated, giving little to no time for preparation. For this reason, it is crucial to invest time in planning and becoming ready well in advance of any crisis.

WHAT IS DISASTER PREPAREDNESS?

The preventive actions done to lessen the intensity of a disaster's impacts are referred to as disaster preparedness. Preparing an organization for an influx of activity, minimizing the impact of disasters on vulnerable groups, and creating a coordinated plan that minimizes resource, time, and effort waste are the three main objectives of disaster preparedness. Disaster preparedness attempts to restore normalcy to the affected populations as soon as possible and has the ability to preserve the greatest number of lives and property during a disaster.

Disaster preparedness is viewed as a "continuous and integrated process comprising a wide range of actions and resources from multi-sectoral sources" by the International Federation of Red Cross and Red Crescent Societies (IFRCRCS). Mitigation strategies, such as infrastructural development and disaster awareness training, are implemented as part of disaster preparedness. However, there are certain distinctions between mitigation and preparedness.

Mitigation efforts by themselves are insufficient to avert a calamity. Mitigation strategies lead to decreased susceptibility. It entails ensuring that the neighborhood and emergency services are ready to handle real-world calamities. Any action that helps increase one's ability to handle a calamity falls under the umbrella of disaster preparedness. Creating and implementing an action plan is the general idea of preparedness.

WHAT IS DEEMED AS AN EMERGENCY OR DISASTER?

When the potential for catastrophic health effects overwhelms a population in an uncommon circumstance, a situation is declared an emergency or catastrophe.

- Bioterrorism
- Chemical emergencies
- Radiation emergencies
- Mass casualties
- Disasters and severe weather
- Outbreaks and incidents

It takes work and time to get ready for a calamity. Planning is necessary so that you can:

- Avoid injury and help others.
- Minimize damage to your property.
- Be alive for at least 72 hours following a calamity at your house or place of employment without assistance from emergency response personnel.

WAYS TO PREPARE FOR A DISASTER

Although it may seem challenging or time-consuming, being prepared is actually extremely easy with a little assistance.

- Get a Kit: Find out what should be in your family's survival kit.
- Make a Plan: Make sure you and your family have a solid emergency plan in place.
- Be Informed: Recognize the catastrophes that are most likely to strike your region and the safety precautions you need to take.

MORE WAYS TO PREPARE

- Download an survival Apps
- Learn Hands-Only CPR
- Make a First Aid Kit
- Print out the Be Red Cross Ready Emergency Contact Card.

Any calamity that strikes—be it a fire, storm, flood, tornado, or act of domestic terrorism—has the capacity to spread chaos and confusion. A large portion of the disruption caused by the unanticipated catastrophe is mitigated by effective disaster preparedness. Having a documented plan in place is essential. Regular review of disaster plans is necessary to make sure everyone understands them completely.

WHAT IS THE SURVIVAL/PREPAREDNESS MINDSET?

We may either become what we think we are or not. The most crucial aspect of our preparedness is our mental state. No matter how much equipment you have, if your mind is weak, you will not survive since it is not capable of handling the kinds of conditions you would encounter.

A preparedness attitude essentially entails training your mind to withstand the stress that an emergency or disaster will bring.

How can I prepare my mind to deal with crises and natural disasters?

1. Be aware. Developing the habit of meditation can help you become more self-conscious and aware of your environment. You may have a deeper awareness of your body, your reactions, and the world around you by practicing meditation. If you're not a meditation practitioner, try practicing generally being more aware of your behaviors and responses. In addition to paying closer attention to every tiny detail of what is happening around you. Look for exits, visit a coffee shop and people watch (in a non-scary way, of course), and truly take some time to disconnect from electronics and observe your surroundings. Self-awareness and situational awareness will become more and more automatic with practice.

2. Learn all the skills. Although I think gear is awesome, we also need to know how to function without the stuff we adore. Every now and again, get outside and force your mind to see beyond the equipment on which we have grown used. Spend a day outside with just a knife and consider how

you may be able to live with it alone. You'll be able to consider the possibilities of changing and succeeding. Never give up learning.

3. Train with your gear. As preppers, we have gear that we love and use for preparation, therefore we must practice with it even though we need to think creatively and step away from it. As much as you can, train using your equipment. Know the equipment as well as you can. Make use of your preparations when trekking and camping as well as at home.

4. Learn from failure. Admitting failure is something that a lot of individuals struggle with since they aspire to be as amazing as their heroes. However, what's the deal? Before they became the fearsome figures you see today, your heroes went through a lot of hardship. They probably still suffer now, as I would wager. You know what, though, makes them fierce? recognizing their errors, taking responsibility for them, and improving oneself via self-improvement. Admit your errors. Take advice from them. Continue trying.

5. Don't have a victim mindset. Saying phrases like "I can't" leads to immediate failure. Your behaviors are dictated by your mentality. Saying "I can't" all the time won't get you there! Saying "I can" will lead you to discover a solution! Your remarks have great impact. Generally speaking, there is always a fix. Be upbeat; it works, I promise. Inhale deeply and consider the resolution. Clear your own path. Give up whining, stop playing the victim, and take control of your own destiny. You are capable of and WILL succeed! You'll go to any lengths to see tomorrow. Never give up! You'll get to the top by crawling and scratching!

6. Get out of your comfort zone. Living in a civilized community means having a roof over your head, air conditioning and heating at your fingertips, Netflix available whenever you want it, and easy access to the internet with just a twist of your wrist. Since we do live in a civilized society, there is nothing wrong with that. Even if a lot of us struggle to make ends meet, having a job is still advantageous to other people. I'm trying to say that we feel at ease. Step out from your comfort zone now. On the coldest night of the year, go camping. Take a hike when it's raining. Start a fire with damp items. Take a risk and do something scary. You're not pushing yourself to see what you're truly

capable of if you're not stepping outside of your comfort zone at least once a month. You won't realize how much you are capable of unless you take action and give it a go! I know it might be unsettling to step outside of our comfort zones. But you have to. Show yourself that you are capable of surviving and navigating difficult situations!

7. Keep your mind open. When someone tells me that they're "doing it incorrectly," I find it really annoying since they're not actually doing anything incorrect; they're simply doing it in a different way. Perhaps that's not how YOU would do it, but what's the issue if it achieved the same outcome with little to no additional work? Be open to adopting new perspectives and ways of doing or thinking about things. I'm not suggesting you have to agree with everyone; it's acceptable to have different opinions. I'm trying to convey that you should pay attention because you never know when they could show you something or open your mind to a completely different way of thinking. Be open-minded, give it some thought, and decide for yourself whether or not this is the appropriate move.

8. Be flexible. When things don't work out the way you planned, accept that bad things happen. Proceed. In an emergency or disaster, you'll need to swiftly adapt and overcome, so how can you be flexible in the ever-changing environment of an emergency or disaster if you can't be flexible in the fluctuations of everyday life?

9. Keep up morale. The best thing to do is to carry some kind of morale item to keep your mind focused and your spirits up since you will become stressed out very soon. Gum can help me stay focused, offers me something to do, and lifts my mood, so it can be as easy as that. It may also be a motivational patch with a quote that inspires you to keep moving forward. It may be a notebook that eases some of your anxiety-inducing thoughts. It may be a book. It may be a card game. It may be an image. Whatever it is, make sure it will actually make you grin and maintain your positive attitude. You'll need something to lift your spirits when things get hard.

10. Be goal oriented and celebrate your goals. Setting objectives might aid with mental focus. If you're walking six miles and you're becoming weary, set little objectives for yourself to help you finish. "I'm going to take a rest at the shade up there," for instance. You cheer when you find shade and can take

a rest! For instance, set a goal for yourself this week to purchase two more cans of food for your food store. Celebrate your success once you've achieved it! Big goals aren't necessary; in fact, the more reachable and reasonable they are, the better! It will be simpler to achieve the larger goals the more you succeed in achieving the smaller ones. In actuality, you will be achieving your larger goals as you attain each smaller one. Most individuals won't buy two months' worth of food at once; instead, they'll gradually stockpile it until they reach their final objective. Set realistic objectives that you are confident you can achieve. Honor each and every objective you've accomplished, regardless of how large or tiny.

The idea of preparedness is something that never goes away. Instead of attempting to do everything at once, focus on one task each week to strengthen your mental toughness and increase your capacity for adaptation and conquering life's challenges, both in normal circumstances and during crises and natural catastrophes.

BOOK 2: PREPPING FUNDAMENTALS

Prepping is a way of life, not just a pastime. Chaos can strike at any time and cause havoc with your daily schedule. When this happens, a lot of individuals worry and try to get ready for the worst-case scenario after it has already happened.

Those that aren't panicked, meanwhile, have made the appropriate preparations for these occasions far in advance. Discover how to begin preparing for newcomers without being intimidated or perplexed.

We'll go over how to begin the process of becoming a prep in this tutorial. We'll go over the essential beginner prepper checklist and how to prepare for several natural catastrophes that might affect your livelihood based on where you live. For your convenience, we now provide a downloaded version of this survival guide.

How to Start Prepping for Beginners

As a newbie to the realm of preparation, you have to lay a strong basis first. This implies that you must understand emergencies and be ready for them. Certain scenarios will always apply to all preparedness enthusiasts.

Additional scenarios will be more tailored to your region and the possible things that might happen there. For example, the likelihood of wildfires posing a hazard is higher on the West Coast region than it is in the Midwest.

If you live in a coastal state like Florida, the same rules apply. The probability of meeting a hurricane is considerably greater than that of a snowstorm. Here are six essential actions that will get you started on the appropriate preparation road as a novice.

The Beginner Prepper Checklist

- Establish a strong base.
- Prepare your house for two weeks of independence.
- The ability to quickly evacuate your house ("bug out bag list").
- Get home bags and take them with you every day to be prepared for situations that may arise while you are away from home.

- Gain basic survival knowledge and practice using your equipment.
- While learning new things and expanding on what you already know, share and recruit.

How to Prep a Solid Foundation

If you're new to the realm of preparation, the first thing you should do is set a solid foundation. As a prepper, you should start by assembling a small supply of basic necessities.

These products are frequently referred to as evergreens. Water reserves, emergency food supplies, first aid kits, hygiene goods, and sanitation supplies are examples of evergreen products. These are the first things that each new prepper ought to stock up on.

FOOD PREPPING

Although food may be kept for longer with either method, the completeness of moisture removal in freeze-drying makes the shelf life of the product greater. Because of its longer shelf life, freeze-dried food is a popular option for emergency food storage, especially when long-term plans are being made. To extend the freeze-dried food shelf life of dehydrated food and help it achieve the stable 25-year freeze-dried food shelf life, you can (and we do!) employ high-barrier packaging and oxygen absorbers.

In an emergency, emergency food might be the difference between feeling calm and panicked. As a result, Valley Food Storage advises that the first step in novice preparation be stockpiling at least two weeks' worth of emergency food. More experienced preppers frequently stockpile food for six to twenty-four months.

However, the more emergency food you have on hand, the longer it will last, and you'll be more prepared. Food for emergencies, such as freeze-dried meats, fruits, and vegetables, should be available.

For survival, knowing how much food to prepare in advance is essential. Depending on factors including age, sex, activity level, and body type, an individual's calorie requirements for health maintenance vary.

It is important to keep your energy levels up in any survival circumstance. It's critical to organize your emergency food supplies according to your household's unique nutritional needs. For example, meals strong in protein and energy are particularly useful during a crisis since they provide long-lasting energy and are frequently a good source of vital nutrients.

Children and younger people usually need more calories since their bodies are still developing. You need to eat less calories as you become older.

Men tend to consume more calories than women do on average. This is due to the fact that men typically have bigger bodies than women do. How active you are will also influence how many calories you should eat.

Generally speaking, you should eat more calories the more active you are. For instance, you will require more food to sustain yourself in a bug-out scenario than in a bug-in one.

Water Prepping

Water is the next item that every novice prepper needs to stockpile. Water storage should be prioritized just as much as emergency food storage. When disaster hits, you won't survive for very long without food or water.

As food and water are the first two items to run out in stores, keeping your own supply on hand is essential for efficient survival planning. For instance, you must keep two different types of water when you first start your water supply.

A minimum of two weeks' worth of clean drinking water should always come first. This entails storing 14 gallons of water. This is based on the average amount of water that people consume each day. The second kind of water you should store is collected rainwater. A wonderful approach to store a lot of water in rainwater barrels is to collect rainwater.

It's crucial to keep in mind that your survival kit should include a dependable filter system or pills for purifying water while thinking about how to begin your preparations. This guarantees that you may utilize your rainwater storage to its fullest potential, even drinking it when it has been adequately cleaned.

The first application of rainwater is drinking water purification. Once your usual supply of emergency water runs out, you should take this action. Reusing collected rainwater to replace flowing water is its second purpose. Rainwater collected, for instance, is excellent for cooking, cleaning, watering the garden, laundry, and other home tasks.

One of the first things to break when there is an infrastructure collapse is flowing water. In this blog post about how rainwater collection might aid in survival, you can find out more about gathering, processing, and cleaning rainwater.

FIRST AID PREPPING

First aid kits are another timeless item that is necessary for novices. Hospitals frequently experience overcrowding and longer wait times during crises. You must be able to take care of your injuries on your own while you wait for medical attention when this occurs. This include being able to splint fractured bones or heal injuries to prevent infection. The Red Cross has suggested the first aid materials listed below.

- Bandages
- Antiseptic wipes
- Adhesive tape
- Adhesive bandages
- Antibiotic ointment
- Gauze pads
- Leg splint
- Arm splint
- Latex gloves
- Emergency blanket
- First aid guide

The requirement for prescription medication is one facet of survival preparation that is frequently disregarded. Keeping an emergency supply of prescription medications on hand might save your life if you or any family member is dependent on them. Consult your physician about obtaining an emergency supply of any prescribed drugs.

In addition to gathering the supplies you'll need, it's a good idea to acquire the necessary abilities. It's likely that seasoned preppers with years of experience already possess the first aid knowledge needed to administer medical treatment in an emergency. But novice preppers might not. Because of this, registering for Red Cross basic first aid and CPR training might make you a hero when seconds count.

Prepping Hygiene and Sanitation

Those who are just starting out should stock up on hygiene and sanitation items. These things are whatever you use to maintain a clean house or good personal hygiene.

Keeping oneself clean during a crisis might help avoid a lot of health problems. Don't forget to incorporate supplies like wet wipes and hand sanitizers into your survival plan. When water is limited, they can be used to stop the spread of illness and pathogens.

Depending on the kind of emergency, certain goods may be difficult to locate or get.

As a result, we advise maintaining a supply of every hygiene item you use in your restroom. These consist of hand soap, dental care, feminine hygiene products, hair products, and toiletries.

This will lessen the likelihood that your health may deteriorate. We also think that having a supply of cleaning materials on hand is essential for novices.

Your preparedness survival pack should have disposable gloves and waste bags in addition to standard cleaning materials. These will support trash management, which is an essential part of starting the preparation process, in addition to helping to keep things clean.

Maintaining a stock of cleaning supplies is essential to preventing the spread of illness. The likelihood of disease spreading increases with the cleanliness of your house. Furthermore, as we have already discussed, hospitals are overflowing during catastrophes.

Maintaining a clean home keeps you away from crowded hospitals and reduces your risk of illness.

Prepping Survival

Reestablishing daily life after a natural disaster might take weeks or even months. For this reason, we advise you to prepare yourself so that you can be self-sufficient for at least two weeks, if not longer. For instance, you must have a strategy to obtain extra food if your food supplies won't last for several weeks. Hunting, fishing, trapping, and food harvesting are all necessary abilities to maintain independence.

When it comes to food storage, Valley Food Storage offers an extensive selection of emergency food solutions that may literally save lives in these situations. These foods guarantee that you have a consistent supply of food, which lessens the need to hunt or forage right away. They also have a long shelf life and are simple to prepare.

Similar to this, if the water you have prepared in advance runs out, you must have a strategy in place for how you will continue to collect and purify water. We advise having a minimum of two distinct approaches for gathering and purifying water.

In some cases, bugging out or fleeing could be the best course of action. Consider, for instance, a storm making landfall at your house in a flood-prone location. In this situation, it would be advisable to flee to higher ground and wait out the storm somewhere safer.

You need a bug-out bag in order to prepare yourself while you're on the way and to evacuate effectively. A bug-out bag is often a backpack that is crammed with of various survival essentials that will keep you nourished, watered, and more comfortable throughout your evacuation.

Take these suggestions into account when you start assembling your bug-out packs.

- Every member of the family has to be aware of the whereabouts of their bug-out bag.
- In addition to other survival supplies, bug-out bags should include a supply of clean drinking water and food that is ready to eat.
- Travel-sized and lightweight gear is what you should include in your bug-out bag.
- You should keep your bug-out bags somewhere dry and cold. You might, for instance, store yours in a designated plastic tote in the garage and wrap them in garbage bags.
- Using this list of basics for a bug out pack, get started assembling your own.

How to Prep for Emergencies that Happen Away From Home

When sh*t is about to hit the fan (SHTF), ideally you'll know it ahead of time. That being said, you can never predict when a disaster may hit. That's why, wherever you are in the world, being ready is crucial.

Consider the scenario when you are going to leave your house to perform errands and an earthquake occurs. You might not be able to get back home soon away in this situation. In this instance, though, you'll be happy that you have emergency supplies ready in your vehicle.

A crucial part of being adequately prepared, especially for those of us who drive to and from work regularly, is packing a bug-out bag for your high-water car.

Here are some suggestions to think about for your bug-out bag, sized for travel.

- Always carry an extra set of clothes in case you are discovered wearing your work clothes.
- Prepare an alternate pair of cozy shoes in case you have to swap out your formal dress shoes, high heels, or sandals.
- In the event that your primary route becomes blocked, think of other ways back to your house or bug-out spot.
- Create a communication strategy in advance with your family in case you are unable to return home soon.

Learn Core Skills, Practice With Your Gear & Maintain Your Equipment To Prep

Not all disasters can be avoided by simply stockpiling food, water, and equipment until your home is well stocked with emergency supplies.

Conversely, possessing the right number of supplies, being aware of where they are, and being able to use them effectively are all necessary for being suitably prepared. Not to mention maintaining them so that no matter how long it has been since you began preparing, your supplies are always ready.

The mental preparation is one part of being a prep that is frequently disregarded. It is important to recognize that catastrophes and emergencies can be emotionally, psychologically, and physically taxing. How well you can handle an emergency scenario may be greatly affected by mentally and emotionally preparing yourself for this stress.

The old saying, "If you give a guy a fish, you feed him for a day," is certainly familiar to all of us. A man is fed for life if you teach him to fish.

This saying captures the spirit of survival preparation. It's not only about assembling materials; it's also about learning and developing abilities. You may be more independent the more knowledge you have.

Learn Core Skills

- Hunting, fishing, trapping, and foraging
- Fire building
- Survival shelter construction
- Water collection and treatment
- Orienteering

Practice With Your Gear

- Cook on a camp stove
- Practice pitching your tent
- Navigate and orienteer with a map and compass
- Grow a garden

Practice Making A Fire Utilizing Different Techniques

- Maintain your Equipment
- Check for expiration dates on your food and medications
- Label and organize your emergency supplies
- Replace dead batteries
- Store your supplies in an area that is protected from water damage and rodents

Connect and Share Your Prepping Experience With Other Preppers

It might feel like you're alone in the world while you're preparing for calamities and disasters. That, however, is completely untrue. As it happens, there is a whole community and network of like-minded preppers out there who, like you, are enthusiastic about safeguarding their family and staying independent.

The internet is a treasure trove of information these days. For instance, there are a ton of useful YouTube channels and websites that may assist you in starting the preparation process. In a similar vein, some preppers choose to collaborate and impart their information to other groups.

Lastly, keep in mind that survival preparation is a process rather than a goal. It's a way of living that requires ongoing education and adjustment. Therefore, if you don't immediately understand everything, don't give up. It's crucial to start someplace and keep getting better with time.

BOOK 3: VARIOUS SCENARIOS TO PREPARE FOR

Disasters can occur in a wide variety of ways, and their duration can range from hours of disturbance to days or weeks of continuous damage. A list of the several kinds of catastrophes that can affect a community, whether they are man-made, natural, or technical in origin, is provided below.

Each year, thousands of individuals are impacted by both natural and man-made disasters. Such major unfavorable events have the capacity to result in a disastrous death toll and physical devastation. They may startle entire populations since they are frequently unanticipated.

Those who witness a disaster may feel distressed emotionally. Anxiety, insomnia, persistent concern, and other depressive-like symptoms are frequent reactions to catastrophes both before and after the occurrence. With the support of their loved ones and the community, many individuals are able to "bounce back" from disasters, but some may require more assistance in order to manage and continue on the road to recovery. Everyone can be at danger, including first responders, recovery workers, and survivors who reside in the affected areas.

NATURAL DISASTERS

Large-scale geological or meteorological phenomena that have the potential to result in property or human casualties are referred to as natural disaster. Among these disaster categories are:

- Agricultural diseases & pests
- Damaging Winds
- Drought and water shortage
- Earthquakes
- Emergency diseases (pandemic influenza)
- Extreme heat
- Floods and flash floods
- Hail
- Hurricanes and tropical storms
- Landslides & debris flow
- Thunderstorms and lighting
- Tornadoes
- Tsunamis
- Wildfire

- Winter and ice storms
- Sinkholes
- Drought

The most frequent natural disasters to be reported in the US are severe storms and floods. Presidential "emergency declarations" may precede these climatic occurrences, necessitating state and municipal preparations ahead of time, including evacuations and the safeguarding of public property. Before, during, and after a natural disaster, Disaster Distress Helpline professionals are ready to chat with callers or text messagers.

HUMAN-CAUSED DISASTERS

Examples include:

- Hazardous materials
- Power service disruption & blackout
- Nuclear power plant and nuclear blast
- Radiological emergencies
- Chemical threat and biological weapons
- Cyber attacks
- Explosion
- Civil unrest

These kinds of traumatic occurrences have the same potential to result in property and human casualties as natural catastrophes. They could also cause populations impacted by them to evacuate particular locations and overcrowd mental health providers.

Other Incidents of Mass Trauma

People can experience intense emotions in response to tragic events, epidemics of infectious diseases, and instability in the society.

PREPPING FOR WILDFIRES: A BEGINNER'S GUIDE

Living on the west coast means that you will probably come across wildfires. Although they are not yet exclusive to the west coast, the majority of these fires occur there. In the realm of preparing, you have to be prepared for the possibility of encountering a wildfire-like disaster. And planning ahead is the greatest approach to be prepared for a disaster.

Preparing your home for a wildfire is one thing to think about. This entails establishing a defensible space—a region with less vegetation—around your home in order to contain the fire and lessen its severity. It is also safer for firemen to protect your property from harm in this area. Normally, the defensible area consists of 70 feet of decreased fuel space beyond the home and 30 feet of "lean, clean, and green" space directly surrounding the house.

Wildfires are difficult to suppress because they spread quickly. You must thus get ready to evacuate if the fires approach your house too closely. This entails having a bug-out or emergency bag prepared and ready to leave.

Enough items should be in this backpack to get you and your family out of your house and into a safe place. That's enough food, drink, first aid supplies, and hygiene items for at least two days.

You should still have a supply of evergreen preparation materials at your emergency location, and these items shouldn't take the place of your emergency food.

You must secure the six Ps in order to be well prepared for a wildfire.

1. People and pets: Make sure that all members of the family, including your priceless pets, are present and accounted for.

2. Phone numbers and important documents: In the event of a wildfire evacuation, staying in touch with doctors, other family members, and emergency services is crucial.

3. Prescription drugs, vitamins, and eyewear: Just because you had to leave your house doesn't mean that your health and wellbeing had to suffer.

4. Images and important keepsakes

5. Personal computers and other essential technology: Maintaining your PCs and other key equipment will enable you to stay online and connected.

Plastic, which includes cash, credit cards, and ATM cards: Having cash, credit cards, or ATM cards after an evacuation will let you make hotel reservations and purchase food and water.

Prepping for Hurricanes

The preparation for a hurricane is similar to that for wildfires for a novice. You should be ready to leave your house at any time in both scenarios. You should prepare a bug-out bag just like you would for a wildfire.

Enough food, water, first aid supplies, and sanitary items should be included in these packs to get you to your emergency location. The ability to adequately prepare your home for a hurricane is the difference between the two tragedies.

Boarding up your windows and doors is one way to get your house ready for a hurricane. Adding reinforcement to your windows and doors can help lessen the amount of water damage that floods inflict to your house. In light of this, we advise maintaining a supply of plywood and nails on hand to cover every window and door in your house. For the sake of survival, it is preferable to be prepared for a flood.

Keeping a supply of sandbags in your house should also be on your hurricane preparation checklist. Greywater backs up in the system due to the amount of water that enters during a strong storm. Because our streams are overflowing, the excess water causes drainage system breakdowns.

Sandbags are an effective way to reduce the quantity of gray water that seeps into your house. You should cover the floor wastes and drains in your house with sandbags. These drains are frequently located in basements, laundry rooms, shower rooms, and bathrooms.

Having a strategy to monitor the storm's progress should be taken into account while preparing for hurricanes. A hand crank or battery-operated radio that receives the NOAA Weather Radio All Hazards network is one example of this. Additionally, make sure you have a list of emergency numbers that includes the numbers for your local emergency management office, police enforcement, fire and rescue services, public safety, electricity companies, and your family members. Gray water backing up into your house and creating water damage may be prevented by utilizing sandbags.

Prepping for Tornadoes: A Beginner's Guide

The next category of natural disaster that can force you to evacuate your house is a tornado. To be as prepared as possible, we advise you to use our tornado emergency pack checklist.

How can one begin tornado preparations? It's easier than you may imagine.

Similar to hurricane preparedness, there are things you can take to make your home safer during a tornado.

To lessen the possibility of threats falling from above, you can trim any dead or broken branches from the trees that surround your house.

It is advised that you clear the area surrounding your yards of any trash and movable objects, such as BBQ grills and wood stacks.

Strengthen the areas where your house is most vulnerable, such the doors, windows, and garage. This is a crucial component of tornado survival planning.

It's critical to understand the distinction between a tornado watch and a warning during tornado season. A tornado watch is a kind of early warning that alerts you to the possibility of a tornado and states that the weather is becoming worse. During a tornado watch, we advise getting your house ready and your bug-out bags ready.

Conversely, a tornado warning is sent out when weather radar detects or indicates the presence of a tornado. Severe tornado warnings frequently suggest that, should the tornado get close enough to your property, you may need to evacuate.

BLIZZARD PREPPING FOR BEGINNERS: HOW TO PREP FOR THE COLD

The last kind of natural disaster for which you may need to prepare is blizzards. People in the northern midwestern states are affected by blizzards. Unlike the other calamities we discussed, blizzards are different.

You should stay inside your house and bunker down, not run out of it. It becomes essential to have a supply of survival food buckets in your home as a result. It becomes essential because you have to have enough supplies to last till the snowfall stops. Additionally, having little goodies on hand, like freeze-dried fruit, might help pass the time during these exhausting days—especially if there are small children around.

To be prepared for blizzards, we advise keeping at least two weeks' worth of supplies on hand. Preparing for blizzard survival in this way might become second nature.

A snowstorm won't persist for two weeks in total, but its effects might certainly linger for two weeks. These consequences include unsafe road conditions, vacant grocery stores, and breakdowns in the infrastructure. You can leave your house and replenish your supplies when the snow passes.

Power Outages

Similar to blizzards, power outages might force you to spend a lot of time holed up in your house. Power outages can be caused by blizzards, but they can also occur from other sources.

Make sure you have adequate supplies for two weeks in case of a power outage. Make sure you have a strategy in place for both restoring electricity and emergency illumination while getting ready for a power outage.

Here are some survival techniques for when there are power outages.

- Get additional gasoline and an emergency generator ready in advance.
- Use solar panels to power gadgets such as radios, headlights, and mobile phones.
- Maintain an assortment of batteries in various sizes.
- To keep your vital gadgets powered, use rechargeable batteries and auxiliary power banks.

In the event that your cooking depends on power, you should also make sure you have a means of cooking your emergency food. Purchasing a camping stove, storing propane and other fuels, or learning how to cook without power are some ways to do this.

The Essential Beginner Prepper Checklist

The materials you'll need are the same for each of these scenarios, but how you use them will vary. We have put up a beginner's prep list to make things easier for you.

No matter what, you should be stockpiling these things. Of course, there are other things you should stockpile in addition to the items on our list for novice preppers. However, it's a fantastic place to start.

- Emergency food that lasts 25 years
- Emergency water

- First aid items
- Emergency hygiene items
- Emergency cleaning products
- Emergency materials (plywood, nails, sandbags)
- 72-Hour Emergency Food Kit
- Family Bug out bag Electronic charges/backup communication
- Emergency long term gas storage for best bug-out vehicle

Well done for starting your prep adventure! Time, effort, and a substantial financial outlay are all necessary for prepping. However, when you can keep yourself and your family more comfortable in the wake of an emergency situation or unanticipated disaster, all the effort you put into the process will be worthwhile.

BOOK 4: CRISIS PREPAREDNESS GUIDE

A crisis can strike at any time, and if a business, educational institution, government agency, or healthcare facility is unprepared, the consequences might be disastrous. Conventional crisis management tactics were mostly concerned with dealing with natural catastrophes like earthquakes, fires, and extreme weather. Administrators and incident managers, however, also need to be ready for a wide range of additional crisis scenarios that may arise in the modern world, such as international disease outbreaks, cyberattacks, supply chain disruptions, power and IT failures, operational mishaps, and active shooter incidents.

A crisis can have a wide range of serious effects. For instance, natural catastrophes can result in expensive damage to infrastructure and institutions. The National Centers for Environmental Information estimates that since 1980, weather-related disasters have cost the United States $2.275 trillion. Epidemics can sometimes have disastrous effects. It could take years for the global economy and supply systems to recover from the devastating effects of the coronavirus epidemic.

Violence-related incidents have a substantial negative psychological and physical impact on victims as well as the community at large. During and during a crisis, people may go through a variety of experiences, including psychological suffering, bodily harm, and emotional pain. There is even data to show that K–12 pupils' mental health is adversely affected by active shooter training.

Crisis circumstances are obviously irreversible. It is imperative that employers, government representatives, educators, and healthcare facilities create all-encompassing crisis management plans. These plans need to be more than just a list of things to do in case of an emergency. A crisis management plan has to contain actions to both completely and promptly recover from a crisis as well as to minimize and prevent them where feasible.

THE FOUR STAGES OF A CRISIS

Pre-crisis, crisis, reaction, and post-crisis are the four distinct phases that usually characterize a crisis. According to some sources, there are up to six stages of a crisis. These stages include mitigation and preparedness in the pre-crisis stage and recovery and learning in the post-crisis stage. Notwithstanding these subtle variations, the fundamental idea is still the same: crises (and crisis management) consist of several components, all of which need to be taken care of in a complete plan.

PRE-CRISIS PHASE

The period of time leading up to a disaster or emergency occurrence is defined by this initial crisis stage. In the pre-crisis stage of some crises (like a cyberattack), there may be no warning indications that something bad is about to happen. In other situations, warning indicators could exist in the lead-up to a crisis. For instance, meteorologists are usually able to predict when a hurricane will make landfall and issue warnings. Sometimes warning indicators, behavioral or psychological, are displayed by an active shooter before they carry out their assault.

Administrators, emergency responders, and leaders may become ready for a crisis during the pre-crisis phase. Pre-crisis activities usually include preparedness and mitigation.

CRISIS PHASE

This is also referred to as the acute phase and is when the crisis becomes visible. The event cannot be stopped at this point; it will occur. Thus, during this stage, incident managers and administrators need to concentrate on three tasks:

Assess: Learn as much as you can about the circumstances. What is the emergency, disaster, or threat? Who is in danger? What kind of reaction is needed?

Activate: The next step is to initiate the appropriate response plan based on the responses to the assessment questions.

Alert: Certain elements of a response plan are situation-specific (e.g., responding to a cyberattack as opposed to a violent physical attack), while other elements are standard. Incident management need to have a solid strategy and procedure in place for alerting impacted parties and providing them with instructions on what to do in the event of a crisis.

Emergency communication is crucial during any crisis. As the Federal Emergency Management Agency (FEMA) points out, in an emergency, communication is essential. "Well-crafted and well delivered emergency communications may support response operations, safeguard property, guarantee public safety, foster collaboration, build public trust, and assist in the reunification of families."

An emergency notification system has to give administrators the ability to swiftly and through various channels connect with targeted audiences in order to be effective in today's society. Corporate communications solutions ought to work with the alarm and notification systems that are already in place.

RESPONSE PHASE

When incident administrators select a course of action and provide resources to counter the danger, this phase begins. This phase might last for a wide range of times. This phase might last for many hours in the event of an active shooter situation or several days in the event of a natural disaster. In some situations, the reaction period could last considerably longer; for example, numerous organizations are still addressing the COVID-19 outbreak almost two years after it first started.

POST-CRISIS PHASE

During this stage, crisis management is abandoned and regular activities are resumed. Government agencies may resume their regular daily activities, employees may return to work, and schools may reopen for business. There is more to this phase than merely going back to "business as usual." Injuries can need medical attention, and facilities might need to be fixed.

The crisis's effects on mental health must also be addressed throughout the settlement stage. This might entail starting status checks with those who are impacted and/or giving them access to options like therapy. In order to address widespread anxiety and trauma, incident managers might also need to speak to the community as a whole and offer tools and information.

In order to determine what happened well and poorly, the post-crisis phase should also concentrate on event analysis. Employers, government organizations, and incident managers may all benefit greatly from using after-action reports to refine their crisis management plans in advance of the next occurrence.

BUILDING A COMPREHENSIVE CRISIS MANAGEMENT PLAN

More than merely crisis management procedures are covered in a comprehensive event management strategy. Every phase of an event—before, during, and after—needs to have a strategy.

PROACTIVE

The main goal of proactive crisis management is to prepare for unexpected events in advance. Mitigation and preparedness are the two components. Although it is impossible to stop every crisis from occurring, there are steps that may be taken to lessen the likelihood of one and lessen the damage that it does to people and property.

Depending on the crisis, different prevention and mitigation techniques apply. While certain natural disasters, like tornadoes and blizzards, cannot be avoided, careful planning may lessen the chance of property damage and worker casualties.

It could be able to stop a crisis in some circumstances. Robust cybersecurity protocols, for instance, might discourage an intruder. Proactive mental health treatment might save someone from committing a violent crime.

The type of crisis may determine different preparation strategies. For instance, businesses and educational institutions may choose to assemble first aid kits and provide CPR training to certain staff members. Establishing emergency response protocols may help organizations be ready for both fires and natural catastrophes. Fire drills are conducted at schools to test alarm systems and response times. Businesses may simulate active shooter situations to instruct staff members on the fundamentals of the Run, Hide, Fight strategy.

Proactive crisis management aims to reduce the likelihood of a crisis (if at all feasible), plan a suitable response, and make sure that everyone involved is aware of how to handle an emergency.

RESPONSIVE

The main goal of responsive crisis management is to respond to a crisis while minimizing disruption to operations and safeguarding persons and property. In certain situations, responding to a crisis may require working with local police enforcement and/or 9-1-1 dispatch.

A strategy and system are also required by incident management in order to safely and promptly notify the impacted parties of the issue. For instance, informing both remote and in-office employees of a cybersecurity breach would be crucial, therefore an emergency notification system would need to have the capacity to concurrently deliver messages across a number of channels (such as phone, SMS text, desktop alerts, and mobile push notifications).

There may be easier response protocols in other situations. In the event of a broken water pipe in a campus building, for instance, it would be imperative to notify staff and students of the facility's closure as well as the cancellation or rescheduling of courses; nevertheless, law enforcement would probably not be required.

Recovery

A crisis management plan's recovery phase should concentrate on treating any physical harm and offering impacted parties financial, emotional, and mental assistance. The psychological consequences of a crisis on both people and groups, for instance, are covered in the Centers for Disease Control (CDC) handbook to Crisis and Emergency Risk Communication. It's critical that incident administrators comprehend these mental health concerns and have a strategy to deal with them.

A procedure for assessing the occurrence and applying fresh information to enhance and broaden crisis preparedness tactics ought to be part of the rehabilitation phase. Finding security holes, weak points in emergency communication, and problems with first response coordination are all possible steps in the learning process.

Emergency Evacuation Plan

You should consider an evacuation plan well in advance of any disaster or unanticipated incident that could require you and your family to leave your house suddenly. From storms with enough warning to a more pressing crisis like a fire, emergencies can take many different shapes and require different amounts of planning time.

A well-thought-out evacuation plan that you and your family members communicate with well in advance is a smart move for overall safety and success in the event of a disaster. Think about who will know where you are, how you will remain in contact, and where you are going and how you will get there.

Step 1: Decide on a location for the family members to gather, making sure it is beyond the affected evacuation zone.

Step 2: Draw out your main escape route, along with backup plans in case the one you originally planned is blocked.

Step 3: Draft a communication strategy to be used in the event that family members split up. Create a backup plan that works for everyone in the event that mobile or landline service is unavailable. Keep in mind that in some situations, public safety representatives will use a variety of channels to inform the public—such as news outlets, social media, and smartphone alert broadcasts—about updates and the necessity to evacuate. In the event that any family members separate, these might be very helpful informational tools.

Step 4: Make sure your car has enough gasoline to go to the meeting spot, keeping in mind that you might not be able to use your favorite escape route.

Step 5: Choose a contact person who is not in the impacted region, and provide that person's contact details to all members of the family so that they may function as a point of contact in the event that you become separated.

Step 6: Text each other using your mobile phone numbers. If your cell reception is bad, text messages will usually still go through.

Step 7: Make sure your survival kit has a portable radio, a charger for your phone, a charger for any tablets or computers, and new batteries so you can access the most recent information. Remember to pack any necessary prescriptions.

BOOK 5: MENTAL HEALTH SURVIVAL GUIDE

For everyone concerned, disasters are distressing events. Sometimes the emotional toll a disaster leaves behind can be even more severe than the financial difficulties caused by physical damage and loss of personal, commercial, or household goods.

Particularly vulnerable groups include young individuals, the elderly, those with functional or access requirements, and those whose first language is not English. Youngsters may experience fear, and some older adults may initially appear confused. Individuals who have functional or access requirements might need more help.

If you or someone in your family is struggling with stress connected to a disaster, consider seeking crisis therapy.

UNDERSTAND DISASTER EVENTS

Recognize each person's own impact from a disaster.

- A disaster affects everyone who witnesses or encounters it in some manner.
- It's common to worry for your personal safety as well as the safety of your loved ones and close friends.
- Feelings of extreme sadness, despair, and rage are typical responses to an unusual occurrence.
- Expressing your emotions aids in your healing.
- Resolving to concentrate on your advantages promotes healing.
- It's healthy to accept assistance from community resources and initiatives.
- Everybody's requirements and coping mechanisms are unique.
- It's normal to desire to get even with those who have hurt you deeply.

Following a disaster, elderly people and children should be given extra consideration. People who witness a disaster "second hand" through significant media coverage may also be impacted.

For therapy, get in touch with your neighborhood's volunteer organizations, faith-based groups, or licensed counselors. Additionally, crisis counseling services may be offered by FEMA as well as the state and municipal governments in the impacted region.

In order to be prepared in case a disaster strikes again, it is a good idea to update your family's disaster plan and restock necessary supplies while you recuperate. Knowing that you are ready for anything will make you feel better all the time.

RECOGNIZE SIGNS OF DISASTER-RELATED STRESS

Adults who exhibit any of the following symptoms may require crisis intervention or help with stress management:

- Difficulty communicating thoughts.
- Difficulty sleeping.
- Difficulty maintaining balance in their lives.
- Low threshold of frustration.
- Increased use of drugs/alcohol.
- Limited attention span.
- Poor work performance.
- Headaches/stomach problems.
- Tunnel vision/muffled hearing.
- Colds or flu-like symptoms.
- Disorientation or confusion.
- Difficulty concentrating.
- Reluctance to leave home.
- Depression, sadness.
- Feelings of hopelessness.
- Mood-swings and easy bouts of crying.
- Overwhelming guilt and self-doubt.
- Fear of crowds, strangers, or being alone.

EASING STRESS

For professional assistance with stress associated to disasters, speak with someone.

Here are some strategies to reduce stress associated with disasters:

- Even though it might be challenging, talk to someone about your emotions, including grief, rage, and other sensations.
- Consult with seasoned counselors who specialize in handling stress following disasters.
- Refrain from blaming yourself for the unfortunate incident or being angry because you believe you are unable to immediately assist with the rescue efforts.
- Make appropriate food choices, get enough sleep, exercise, meditate, and practice relaxation to support your own physical and mental healing.
- Keep up a regular family and daily schedule, and try not to put too much pressure on your family or yourself.
- Get together with loved ones and friends.

- Take part in memorial services.
- Make use of the family, friend, and religious support networks that are already in place.

By replenishing your emergency supply packs and revising your family's disaster plan, you can make sure you are prepared for any eventuality. It might be consoling to carry out these constructive deeds.

HELPING KIDS COPE WITH DISASTER

Disasters can cause youngsters to feel uncertain, scared, and bewildered. It is crucial for parents and teachers to be aware of trauma and prepared to support their children in the event that they develop stress reactions. This is true whether the kid has experienced the trauma firsthand, has only watched the incident on television, or has heard adults talk about it.

When a disaster strikes, kids may react with grief, anxiety, or behavioral issues. Younger kids could revert to past behavioral patterns such bedwetting, insomnia, and separation anxiety. Children who are older may also exhibit violence, hostility, difficulties in school, or withdrawal. Certain youngsters who observe the disaster on television but have only indirect experience with it may get distressed.

RECOGNIZE RISK FACTORS

Many children's responses to disasters are fleeting and typical of their responses to "abnormal experiences." Three main risk factors may put a lower percentage of kids at risk for psychological discomfort that lasts longer:

Direct experience with the disaster, such as being evacuated, witnessing other people's injuries or deaths, or getting hurt while feeling as though one's own life is in jeopardy.

Grief/loss: This refers to the passing away or severe injuries sustained by friends or relatives.

continuous stress brought on by the disaster's side consequences, such as having to move temporarily, losing friends and social networks, losing personal belongings, parental unemployment, and recovery expenses to get the family back to their pre-disaster situation.

VULNERABILITIES IN CHILDREN

Most of the time, unpleasant reactions are transient and contingent on the risk factors mentioned above. Symptoms often go away over time if there isn't a serious risk to one's life, an accident, the death of a loved one, or other secondary issues like moving or losing one's house. Reminders of the disaster, such as strong winds, smoke, overcast skies, sirens, or other similar objects, may bring back unpleasant memories for people who were there firsthand. These emotions may be exacerbated by a past experience of extreme stress or trauma of some kind.

Parents' coping mechanisms are often linked to how well their children handle crises or disasters. They are able to sense grownups' melancholy and worries. By managing their own emotions and coping mechanisms, parents and other adults can save children from experiencing the stress of catastrophes as much as possible. In most cases, parents are a child's best source of support during a disaster. Preparing a family disaster plan with the kids is a great approach to give them a sense of control and confidence before a disaster strikes. Children can help with a family's post-disaster recovery strategy.

MEETING THE CHILD'S EMOTIONAL NEEDS

Adult behaviors, attitudes, and ideas have an impact on children's responses. Children and teenagers should be encouraged by adults to express their feelings and ideas regarding the event. By listening to children's worries and providing information, you may clear up misconceptions around risk and danger. Remain composed by acknowledging the worries and perspectives of the kids and by talking about specific safety measures.

Pay attention to the child's words. When a little child asks questions regarding the incident, simply respond to them without going into the same detail that an older child or adult would need. Determine the degree of knowledge your own kid requires. Some children find comfort in knowing more or less than others. Give a youngster the opportunity to illustrate their emotions or narrate a narrative if they are having trouble expressing themselves.

Make an effort to determine the source of your concerns and anxiety. Recognize that children's main fears after a disaster are:

- The incident will recur.

- They will lose someone important to them or hurt someone else.
- Either the family will split up or they will be left on their own.

REASSURING CHILDREN AFTER A DISASTER

Among the ideas to assist comfort kids are the following:

- Making personal touch is comforting. Give your kids hugs and touches.
- Calmly give accurate details on the most recent disaster, as well as the current safety measures and recovery strategies.
- Encourage your kids to share their emotions with you.
- Spend more time with your kids, especially before bed.
- Get back into the pattern you used to have for work, school, play, eating, and sleeping.
- Involve your kids by assigning them specialized tasks so they may feel like they're making a difference in the family and community.
- Honor and acknowledge conscientious conduct.
- Recognize that your kids' responses to calamities will vary.
- Invite your kids to assist in updating the family's disaster plan.

It might be appropriate to speak with a professional if you have attempted to establish a comforting environment for your child by following the above steps, but they still show signs of stress, if the reactions get worse over time, or if they interfere with everyday behavior at home, at school, or in other relationships. The child's primary care physician, a mental health specialist who specializes in children's issues, or a member of the clergy can all provide you expert assistance.

MONITOR AND LIMIT EXPOSURE TO THE MEDIA

Children may get anxious and fearful as a result of hearing about a disaster on the news. This is especially true in cases of widespread natural catastrophes or acts of terrorism that result in a major loss of life and material damage. Repetition of an incident can lead to the belief that it is happening again, especially in younger children.

Parents should be there to stimulate communication and offer explanations to their children if they are permitted to watch television or use the Internet, where pictures or news about the disaster may be found. This can also entail parents keeping an eye on their children and sensibly limiting their own exposure to material that causes worry.

USE SUPPORT NETWORKS

When kids take the initiative to comprehend and control their own emotions and coping mechanisms, parents support them. They can accomplish this by creating and utilizing networks of friends, family, community organizations and agencies, places of worship, and other resources that are beneficial to that family. In an emergency or when a disaster occurs, parents can create their own special networks of social support to help them cope with the situation and control their responses. Parents will thus be more equipped to help and be more accessible to their children. When things are hard, parents are virtually always the greatest people to help their kids. However, parents must take care of their own needs and make a plan for their own support before they can help their children. Being ready for a disaster gives the family members a chance to recognize that these events are inevitable and to gather the supplies required to fulfill their immediate requirements. Being ready makes a person feel better able to handle situations; this is also true for youngsters.

Below are common reactions in children after a disaster or traumatic event.
birth to the age of two. When pre-verbal children encounter trauma, they lack the language to articulate the experience and their emotions. They are able to remember specific sights, sounds, and smells, though. Traumatized infants may exhibit irritability, cry more frequently than usual, or crave attention and cuddles. When it comes to youngsters this age, parents' coping mechanisms have the largest impact. As kids get older, they could act out parts of the horrific event that happened years ago but was for some reason forgotten in their play.

Ages 3 to 6 make up preschool. When faced with an overwhelming incident, preschoolers typically feel weak and defenseless. Their diminutive stature and advanced age prevent them from being able to defend others or themselves. They thus experience severe anxiety and uneasiness about being apart from their caregivers. Preschoolers don't understand what it means to be permanently lost. They may believe that the effects are temporary or irreversible. Preschoolers' play activities may repeatedly recreate an incident or disaster in the weeks that follow a traumatic event.

7 to 10 years old is the school age. A child of school age can comprehend that loss is permanent. Some kids are very fixated on the specifics of a terrible experience and want to talk about it all the time. The child's academic performance may suffer as a result of this obsession interfering with their ability to concentrate in class. Children may hear false information from others at school. A wide range of emotions may be shown by them, such as melancholy, widespread anxiety or focused worries of the disaster occurring again, remorse for their acts or inaction during the disaster, rage that the event could not have been avoided, or thoughts of acting as the rescuer.

11 to 18 years old is the pre-adolescent to adolescent period. Children get a more complex understanding of the disaster occurrence as they get older. Their answers resemble those of adults more. Adolescents may engage in risky, harmful activities like drug or alcohol abuse, irresponsible driving, and so on. Some people may develop a phobia of leaving their homes and stop doing things they used to do. The goal of adolescence is mostly to venture out into the world. Following a trauma, one's perspective on the world may appear riskier and more unsettling. Teens can experience tremendous feelings and find it difficult to talk about them with others.

PREPARING EMOTIONALLY FOR A DISASTER OR EMERGENCY

Why is it crucial to have emotional readiness in case of calamities or disasters?
A traumatic experience might result in several losses, which can cause worry and concern about the future. Your surroundings may have changed significantly. It might be difficult to adjust to new situations.
You can feel less stressed and anxious if you are emotionally ready for a crisis or disaster. Being able to control your stress on a daily basis will help you get through difficult circumstances. Additionally, it can hasten your healing from trauma and lessen its long-term repercussions.

HOW CAN I BUILD EMOTIONAL WELLNESS?

Maintaining your mental health requires you to learn about and apply appropriate coping mechanisms to get through life's obstacles on a daily basis. These similar abilities will make it easier for you to handle crises.

The following fundamental actions can help you maintain emotional wellness:

Take care of yourself; schedule time for stress-relieving activities like yoga, deep breathing, meditation, and strolling.

Always seek out and implement little, doable adjustments in your life that will improve your ability to deal. Modest lifestyle adjustments can have a significant impact.

Reduce or give up harmful coping mechanisms, such as smoking, consuming alcohol, overeating, or undereating.

Adopt an optimistic outlook on life by telling friends and family one good thing that happened each day.

No matter how busy life gets, always find time for yourself (e.g., read a good book, listen to your favorite music). You'll have more mental stamina to handle hectic schedules if you take some time for yourself.

How can I build a strong support network?

In addition to being crucial in everyday life, strong support systems are crucial in times of crisis and disaster. Make the time to establish and preserve solid support systems in your life.

When you need assistance, ask for it. Taking up assistance from loved ones and friends might foster a community of support.

Participate in events and rituals to foster a sense of community. Add those who hold special meaning for you.

Spend time with friends and relatives who are encouraging. The benefits of socializing with loved ones extend to your mental and physical well-being.

Learn about your neighbors. It's common for your neighbors to react first in an emergency or disaster. It's critical to know who to turn to for support when things get tough. Knowing who might require your assistance in an emergency is also a smart idea.

Participating in volunteer work or joining a local community association are two ways to get to know your neighbors and the neighborhood.

How can I take better care of myself?

You can manage your emotions more effectively if you take good care of your physical health.

Consume wholesome, well-balanced meals in accordance with Canada's Food Guide. Eat regular meals, please. Limit your intake of sugar-rich snacks.

Control long-term disease. Consult your physician or other healthcare provider if you need assistance.

Make time to sleep. Lack of sleep might have an impact on how you handle problems in your daily life. Getting enough sleep is essential for maintaining mental and physical well-being.

Make sure you're getting enough water. Steer clear of sugar- or caffeine-filled beverages. You can stay hydrated by consuming 250 mL (1 cup) of water every two hours while you're awake. It's critical for your health and cognitive function to drink enough water.

Continue to move. You can manage better if you exercise. A fifteen-minute walk might help you relax. Walking induces a calming and stress-relieving chemical release in the brain.

Taking care of practical details ahead of time can help lower stress during an emergency.

Be Prepared

Even in situations where you are unable to change the stress, you can still choose how you interpret and react to it. Making a strategy might give you a sense of control and help you remain composed. This might assist you in making wiser decisions in an emergency or disaster.

How do I start planning?

Imagine that your home is experiencing an emergency (a fire or flood, for example) and that you have to evacuate right away. Which exits from your house are the best? Identify two or more exits from each room. You've begun your emergency plan now that you've written them down.

Ensure that every member of your family is aware of the strategy and what they must do to be safe. Allow your teenagers and kids to assist with the planning.

HELPING YOUR CHILD OR TEEN PREPARE EMOTIONALLY FOR A DISASTER OR EMERGENCY

Every parent dreams of their child never having to experience an emergency or disaster. However, it's possible that your child may experience a traumatic incident at some time. Your child or teen will respond and recover more effectively from a disaster or emergency if you assist them in early development of appropriate emotional coping mechanisms.

A distressing event is anything that makes a child or teen feel physically or emotionally overwhelmed.

A child or teen can feel insecure, scared, or confused after:

- being involved in a disaster or emergency
- hearing others talk about a traumatic event over and over again
- seeing a traumatic event on TV

If a youngster doesn't know what's occurred or why it's altering their routine, they may find it difficult to manage. A youngster is more prone to experience distress if they feel more terrified or powerless. Providing your kid with age-appropriate assistance and advice can help lessen the impact of a disaster or emergency on you as a parent or caregiver. A kid may better manage their response in the event of an emergency by learning good coping mechanisms from an early age.

Building Blocks for COPING Emotionally

Communicate and Connect with Others

Assist your youngster in acquiring vocabulary to express and discuss their emotions. Reassure them that talking to a trustworthy adult about fears is normal and healthy. Provide your teen with strong role models and encouragement in communicating. By doing this, they may be able to create safe havens for them to confide in and form supporting networks of friends and family.

Optimism and a Positive Attitude Help

Organize frequent family dinners and ask everyone to mention one good thing that happened that day. Instruct your youngster to always find the good things in life, particularly when there's nothing more to be gained than the opportunity to grow from the experience. Celebrate your child's accomplishments (letting them do age-appropriate chores around the house and praising their hard work, for example).

Take Part in Community and Family Events to Create Support Systems.

Have your youngster assist you in assembling a home emergency kit. Let your kids assist in creating a family emergency plan. Look for family-friendly activities that you may attend or assist at.

Nurture and Support your Child or Teen's Healthy Emotional Development

Give them the freedom to express themselves. Attend to your child's emotions without passing judgment. Talk to your child in an age-appropriate manner. Express your concern and interest in learning about your child's coping mechanisms with stress.

In an emergency, it's critical to:

- reassure your youngster that they are secure
- Reassure your youngster that you will keep them safe.
- Keep supporting your child in times of fear.

Go for Help

Have a conversation with your youngster about who to call on when they need assistance.

Show your kids how to ask for assistance.

Keep emergency phone numbers in your memory.

Healthy Habits for Your Child

Preschoolers and toddlers require 12 to 14 hours of sleep every night.

Kids from six to twelve require ten to twelve hours of sleep per night.

Make sure your kids get enough time to play and engage in physical activity.

Ensure that your child follows Canada's Food Guide and has a healthy, balanced diet. Eat less high-sugar meals and beverages.

Healthy Habits for Your Teen

Teens require nine to twelve hours of sleep per night.

Set an example for your adolescent and motivate them to exercise.

Teach and encourage your adolescent to form lots of wholesome interactions.

According to Canada's Food Guide, set a good example for your adolescent and urge them to consume a healthy, balanced diet. Eat fewer high-fat, high-sugar meals and less coffee.

For Children and Teens

Limit your exposure to upsetting images, TV shows, or radio broadcasts on a regular or continuous basis.

Watch alongside your kid if they watch or listen to news coverage concerning catastrophes or disasters. After the occurrence, discuss with your child what they witnessed to help them understand what happened.

During upsetting situations, kids want perspective, confidence, and counsel that is suitable for their age.

Every toddler and adolescent has a unique reaction to an emergency or disaster. As you assist your adolescent in learning coping mechanisms for emotions such as dread, guilt, rage, and helplessness, you are contributing to and fortifying your family's resilience in the event of an emergency or disaster.

BOOK 6: FOOD PREPAREDNESS AND PREPPER'S PANTRY

Every prepper wants to be ready for SHTF with a well stocked pantry, but many are unsure of where to begin. Questions like... What sort of food should I store? are better than responses.

For what duration should I prepare?

What is the price of a prep pantry?

Even so, where should I place everything?

proceed to sprout from left to right.

Have you ever struggled to figure out how much beans in cans to purchase on your next grocery shop run or to implement a functional FIFO method? We sympathize with you, friend.

While it's not simple, building a prep pantry from scratch is also not insurmountable. With the correct knowledge, strategy, and mindset, you can quickly construct an effective one.

All the information you want to construct a prepared pantry is provided below:

WHAT KIND OF FOOD SHOULD YOU STORE IN A PREPPER PANTRY?

It's critical to keep in mind that your prep cupboard is different from your ordinary, daily pantry. The goods you keep here are specifically designated for use in times of need and are, to the greatest extent feasible, not used on a regular basis. To begin, stock up on products that will last for three days. Gradually increase the duration of your pantry to two weeks, one month, six months, and so on.

In light of this, what type of food need to you have in your pantry?

These are a few basic standards.

Stock your prep pantry with the following items:

familiar cuisine that you genuinely enjoy consuming. Whoa! You don't have to stock your prep pantry with bad Meals Ready to Eat (MREs) and hardtack. It may be a separate food cache from your main pantry. Instead, you can try to stuff it full of foods that you enjoy eating.

Keep in mind that when SHTF, your pantry may be your sole supply of food. If you are going to be eating a specific type of food for a long time, it better be something you enjoy eating on a regular basis.

Keeping familiar foods on hand not only preserves a modicum of normalcy but also simplifies the process of preparing meals. This is particularly crucial if you have family members with dietary restrictions or finicky eaters among your children. Additionally, storing familiar foods in your prep cabinet reduces the likelihood of food allergies and reactions.

Shelf stable. Keeping a prep pantry can be problematic due to spoilage, so choose foods that can keep for a long period even without refrigeration. These include of non-perishable dry goods like jerky, dried fruits and vegetables, powdered eggs, canned beans, canned soups, pasta, and white rice, to mention a few.

Long-lasting foods are more economical since they are simpler to store and require less frequent rotation than perishables. To guarantee their quality, store them correctly and give them frequent attention.

both high in energy and nutrients. You would need all the power you could muster to survive. You should prepare for heavy manual effort in the event that power and other modern comforts go out, so stock up on meals high in calories and superb in nutrients.

Foods rich in energy include grains like rice, maize, and quinoa, as well as nuts and beans. Keeping an ample supply of vitamins, minerals, and dietary supplements on hand is also beneficial.

Food you can buy in bulk

rice

Over time, buying food in bulk may save you and your family a significant amount of money. For example, 50 kilogram bags of rice or one month's worth of powdered eggs are available. Purchasing canned goods during grocery store sales is another method to gradually but steadily assemble your prep cupboard and make a little more income on the side.

Food that's easy to prepare

You could not have supplies like fuel or electricity when SHTF. This implies that you need to keep foods that can be prepared quickly and with minimal effort. Above all, store food items that you are capable of preparing yourself.

For instance, it could be better if you don't have wheat bags in your cupboard if you don't know how to make bread from scratch.

Remember that you don't need to purchase each of these foods at once. Purchasing an additional can of tuna or a bag of pasta on your next supermarket journey will allow you to gradually stock your cupboard. Purchase a bag of rice or flour and repackage it into containers more appropriate for long-term storage if you have extra money to spare.

The daily caloric consumption of your household can also be taken into account while stocking your prep cupboard. The average person requires 2,000–2,500 kcal per day, however each person's daily energy needs might vary based on their height, weight, and level of daily activity.

Preserving Your Own Food

In addition to purchasing prepared foods from supermarkets and grocers, think about canning your own food. Food preservation may help you consume genuine, even healthy, food when SHTF, save a ton of money, and reduce food waste.

Learning how to preserve food is helpful, particularly if you have extra supplies or harvest from your survival garden.

The following are some at-home methods for food preservation:

Canning. People used to can food all the time before electricity and refrigeration were invented. They used cans to keep their fermented veggies, meat, jams, preserves, and dry goods. Food that has been canned properly can keep for several months.

drying. Meats were preserved in the past by hanging them outside and letting them dry completely. With the use of a dehydrator, you may now dry a wide range of food items, including fruits, vegetables, venison, and pig. Dehydrators may increase the shelf life of your food from a few days to several months, making them an excellent investment for food preppers.

Healing. This is a further method of food preservation that is frequently used with meat products such as sausage, bacon, and ham. In order to remove water from meat and increase its shelf life, curing mostly employs salt, nitrates, and nitrites.

The process of fermentation. Sauerkraut and pickles are made via fermentation.

How About Freeze Dried Food?

While freeze-dried food may not be the most affordable item to include in your prepared food pantry, it does provide some benefits.

It can persist for decades, to start! Food that has been freeze-dried maintains the majority of its nutritional value and flavor, so you won't have to worry about eating cardboard-like food. It's also quite simple to cook. These meals may be easily reconstituted with just hot water, and then served. These days, you can get freeze-dried dishes like lasagna or beef stroganoff, so if you have extra money, you might want to pick up a few for your prep cupboard.

Selecting the Right Place

The location of your prep pantry is important. To prevent rotting, keep rodents away, and maintain the organization of your cache, it needs the proper circumstances.

By promoting the growth of bacteria and fungus, high temperatures and moisture content speed up spoiling, so make sure your site is continuously cold and dry and away from direct heat and humidity. Photodegradation is a chemical process that can occur when exposed to direct light.

The colors, vitamin and nutritional contents, and food packaging are all destroyed by photodegradation. Cover your windows with heavy curtains or shades to keep direct sunlight off of your treasure.

It's also critical to keep in mind that you should pick a location free from frequent temperature swings. The garage is not the best place for a prep pantry since temperature swings may cause just as much harm as direct heat or light.

In order to keep their goods somewhat chilled during a blackout, many preppers choose the coldest place in the house—such as the basement or attic—as their prep pantry.

Having shelves in your pantry area is also advantageous as it will facilitate much simpler and more effective organizing. Here's a short and simple guide on building a sturdy shelf system for your prep pantry:

Building a Prepper Pantry With Limited Space

What happens if you don't have a spare room, attic, or basement? You don't even have space, so how would you store all this food?

For many preppers, a typical obstacle is a lack of room, but a little ingenuity and resourcefulness may go a long way.

One way is to turn any vacant space in your home into a place to store food.

You may build a tiny room with shelves to arrange your things properly if you're feeling crafty. Make use of areas such as the closet beneath the stairs, an old closet, or, if you have one, your crawl space. To the current furniture, add some pull-out shelves and vertical storage solutions. You may install sliding shelves to store cans under your bed.

What Kind of Storage Supplies Do You Need?

Having determined what sort of food to stockpile in your prep pantry and where to hide it, the next step is to choose which storage items would work best for you.

You would need the following storage tools and accessories in addition to the standard cans, totes, and mason jars to help preserve your food and increase its shelf life:

Mylar bags

One of the most popular products for storing food for a long time is a mylar bag. You're seriously missing out if you don't keep at least some of your food in mylar bags.

The same material used to make space blankets is also used to make these metalized baggies, giving them a flexible, light weight without sacrificing strength. They are resistant to breaking or ripping easily because of their great tensile strength. They're also excellent at shielding your food from oxidation and light damage.

There are several sizes and thicknesses of Mylar bags. The most common sizes are the larger 5-gallon bags that may be used to line food buckets and the 1-gallon bags, which often include a resealable mechanism.

When it comes to thickness, 4.3 mils and up is a good starting point. But keep in mind that heavier mylar bags usually cost more, so before you go on a mylar bag buying binge, decide what you want to store.

Repackaging using Mylar bags is another option:

dry products, such as beans and grains.

baking supplies such as sugar, flour, and baking soda.

drugs, particularly those that react negatively to light, such as vitamins and antibiotics.

How to seal mylar bags

As was previously indicated, certain mylar bags are without zips, in which case you will need to use a heat sealer to close them. Alternatively, if money is tight, you might use a regular clothes iron or a hair straightening iron.

The following instructions will help you seal your mylar bags:

- Give yourself a few inches so that the mylar bag can seal; avoid packing it too full.
- Check your seal's heat first. It won't seal correctly if it's not hot enough. If you keep it set too high, your mylar bags will burn through.
- To prevent breaks, ensure that the edges you're sealing rest flat and are free of minute particles or dirt.
- Remember to add an oxygen absorber to your mylar bags before closing them to prevent spoiling.

FOOD BUCKETS

Five-gallon pails are an excellent alternative for storing food. Because they keep food sealed in an airtight atmosphere, they work wonders at shielding food supplies from heat, moisture, and vermin. The lids on these buckets typically have a rubber mallet to firmly fasten it in place. Once the lids are on, they might be difficult to take off, so be sure to carry a lid remover with you.

In other cases, particularly when the buckets are used, the lids may be broken or missing entirely. Regular lids are available individually, or you can upgrade to gamma lids.

Gamma lids, in contrast to ordinary ones, are twist-on and off and fit in most buckets, so you won't need to hammer them into place. They may cost a little more, but they do provide better food security.

As previously indicated, you can line the interior of your food buckets with 5-gallon mylar bags. These containers can hold liquids like vegetable oil as well as food goods including grains, cereals, flour, and sugar. A different approach is to keep several mylar pouches in a single bucket. Just make sure that the plastic in the buckets you're using is food-grade to prevent any dangerous chemicals from leaking into your meals.

The best part about these buckets is their affordability—they can occasionally be found for free. Make sure the food pail you're recycling didn't previously include any non-food items. It is not appropriate to use a bucket that previously held hazardous substances.

To avoid shattering the lids, don't stack these buckets too high when storing them.

Remember to name them appropriately as well!

OXYGEN ABSORBERS

An attractive addition to your food storage toolkit are oxygen absorbers. These little sachets go a long way toward prolonging the shelf life of your food.

The presence of oxygen fosters the growth of fungus and bacteria. In addition, this element degrades tastes, vitamins, and other dietary ingredients; this is why open air causes apples and potatoes to turn black. Absorbers of oxygen prevent this from occurring. The iron powder in these packets draws oxygen out of the atmosphere. Additionally, there is less chance of spoiling when there is less oxygen in the air.

To prolong the shelf life of your dry products, place one or more oxygen absorbers on top of them before sealing them in. Oxygen absorbers are safe to be placed with food.

How to Organize Your Prepper Pantry

Congratulations! You've come this far!

Now you have the correct food, the perfect location, and the appropriate materials. It's time to start arranging now.

A functional prep pantry is built on efficiency. How much supplies you have is irrelevant. All that work will have been in vain if they're all piled into one enormous pile and you can't find what you need.

The following advice will help you get your prep pantry in tip-top shape:

FIFO Method

The general rule of thumb for medium- to long-term food storage is the First In, First Out technique. The basic premise is as follows, which you are undoubtedly already familiar with: whatever is oldest in your cupboard is eaten first.

When adding new things to your pantry, place them at the rear of the shelves so that you can use them in the order that you purchased them.

Remember to write down the dates of their expiration in Sharpie and keep it in a visible place.

Using Can Dispensers

This is a practical approach to set up your prep pantry using the FIFO principle. These dispensers eliminate the need for you to take out and then put back in your cans each time you need something by allowing them to roll down automatically when they're ready to use, as demonstrated in the video above.

You can use this idea to make use of some of your home's cupboard spaces if you're a touch short on room.

Disaster-Proofing Your Shelves

The shelves in your prep pantry should be fastened to the wall if you live in a hurricane or earthquake-prone location to prevent them from falling off with your supplies. To be extra safe and prevent your jars and glasses from spilling on the floor, it is best to have a lovely bar in front of those shelves.

Have a Working System

Because different people's brains function differently, an organizing structure may be a nightmare for some people and a dream for others. Some others categorize their belongings, some sort them alphabetically, and yet others use color coding to identify the materials in their prep cupboard.

Regardless of the system you use, it's critical to remember the following:

Put labels on everything. This is crucial if you're repackaging goods into food buckets or mylar bags. Place the product's name, the manufacturing and expiration dates in a visible location. You can print labels on sticker paper or use a permanent pen.

The products you use the most should be kept in a convenient location. You won't have to go through your whole prep pantry that way.

Place the heaviest things on the lowest shelves. This reduces the possibility that they will fall and hurt someone, making them easier to transport.

Sort and arrange your supplies according to their categories. You may designate a shelf for breakfast products, another for baked goods, etc. To maximize space and reduce clutter, place goods on the shelf in cardboard boxes or square plastic totes.

Combining all of the components into a single food-grade bucket is another fantastic idea. This enables you to allocate supplies based on the meal you plan to eat that day. Include the recipe (or write it down on an index card) so that, in the event that you are unable to cook, anyone in your family can prepare food.

Keep an Inventory of Your Supplies

You must maintain a neat, organized inventory of everything in your prep pantry in addition to using an effective method for using and storing your goods. This facilitates food rotation and allows you to keep everything under control with a quick glance.

Save a Google document or Excel file with the items you purchased that month. Keep track of the products that are about to expire, the amount you've used or want to use in an emergency, and the quantity you still need to buy on your next grocery trip. It's a simple, methodical way to maintain the organization of your prep cupboard.

Ideas: Well, it wasn't all that horrible, was it?

Initially, building a prep pantry may seem like a difficult and costly task, but with the correct resources and knowledge, it can really be rather simple.

Building your prep pantry now is essential. If you start out slowly, it doesn't matter as long as you persevere. Recall that when SHTF, the food supply system will be among the first to collapse. You want to be the prepared person who has everything at home, not one of those men who fight over a can of beans at the grocery store.

BOOK 7: PREPPER'S COOKBOOK

Are you truly ready for anything that might come up? Of all, it's impossible to truly forecast what will happen if SHTF unexpectedly. It's possible that you won't have access to shops or a food store, in which case having backup plans available is essential.

Here are 25 survival food recipes that could one day save your life to help you make sure you're prepared and ready. Come on, let's cook!

Knowledge is power when it comes to emergency preparedness. An essential skill that might keep you and your family alive in the case of a disaster is knowing how to prepare survival foods.

We've put together a list of 25 must-have survival recipes that you should practice preparing right away so that you'll be equipped to handle any situation.

BASIC PIONEER CORNBREAD

This traditional dish may be basic, yet it has an amazing quantity of nutrients in a single serving. It cooks quickly over an open flame or on top of a camp stove with just a few ingredients.

In addition to some protein and other essential vitamins and minerals, cornbread is satisfying and a good source of carbs for energy.

This is an easy recipe to follow.

Ingredients

- 1 c. yellow cornmeal
- 1/2 c. all-purpose flour
- 1 tsp. salt
- 1 tbsp. baking powder
- 1 c. buttermilk
- 1/2 c. milk
- 1 egg
- 1/2 tsp. baking soda
- 1/4 c. plus 2 tablespoons butter or shortening, divided (Ree uses butter)

Directions

Set the oven's temperature to 450°F. Mix the flour, baking powder, cornmeal, and salt in a bowl. Mix everything together.

Pour the milk and buttermilk into a measuring cup, then stir in the egg. Combine using a fork. Stir in the baking soda. Mix the dry ingredients with the milk mixture. Using a fork, stir until well blended. Melt 1/4 of the butter or shortening in a small bowl. Add gradually to the batter, stirring just until incorporated. Melt the remaining 2 tablespoons of butter or shortening over medium heat in an iron skillet. Into the heated skillet, pour the batter. To level the surface, spread. (The batter ought to pop.) Cook for 1 minute on the stovetop, then bake for 20 to 25 minutes, or until golden brown. Crispy edges are a must!

Hardtack

A kind of hard biscuit called "hardtack" gained popularity during the American Civil War. This recipe can supply vital nutrients during a scarcity and has a long shelf life, which is why it makes excellent survival food.

Because it's a very easy, quick bread that doesn't require rising, making hardtack is a rewarding project for both seasoned and novice bakers.

The pieces can be kept in airtight containers until they are ready to be eaten after they are prepared and cooled.

Ingredients:

- Flour
- Water
- Salt

Directions:

The next step is to combine your components into a dough once you've got them all together. You can use an electric mixer, if you have one, or you can do this by hand with a spoon.

When rolling out the dough on a lightly floured surface, you want it to be sufficiently stiff to keep its shape. After your dough reaches this consistency, roll it out to approximately a ¼-inch thickness and use a knife or pizza cutter to cut into little squares or rectangles (the size doesn't matter).

Lastly, puncture each piece several times with a fork and bake at 350°F for 30 minutes on each side, flipping once, or until both sides are crisp and golden brown.

Pemmican

Native Americans created pemmican hundreds of years ago as a method of preserving meat without refrigeration or other preservation methods. For optimal flavor and nutrients, this recipe blends dried meat with fat, nuts, berries, and spices.

When all other choices are exhausted, this is one of the oldest survival recipes still in use today and a superb portable energy source.

Ingredients:

- Lean meat, cut into strips (venison, beef, elk, turkey)
- Dried berries (cranberries or blueberries work well)

Directions:

Drying your meat is the first stage in preparing pemmican. This can be done in a low-temperature oven (175-200°F) or, if one is available, in a food dehydrator. Depending on how much moisture is in the meat before it goes into the oven or dehydrator, this procedure should take six to ten hours.

It's time to ground your beef into a powder once it dries completely. You can use a blender/food processor or a mortar and pestle for this.

When the ground meat is the consistency you want, combine equal portions of rendered fat and ground meat in a bowl, stirring to properly blend.

Then, add as much dried fruit or berries as you like to the mixture (amounts will vary depending on liking), and stir until all of the ingredients are mixed in well.

Using your hands, shape the mixture into small balls; these are what you will eventually consume!

If stored correctly in an airtight container away from direct sunlight and moisture sources, such as humidifiers or water heaters, your handmade pemmican should keep for up to ten years without refrigeration.

NATIVE AMERICAN FRY BREAD

An iconic and delectable delicacy with centuries-old roots to some of the oldest societies on the nation is Native American fried bread. This classic dish, which has been passed down through the years, is made with basic ingredients like flour, baking powder, salt, and water. It is served hot and comforting.

Ingredients:

- 2 cups all-purpose flour
- 1 teaspoon baking powder
- ½ teaspoon salt
- ¼ cup vegetable oil
- ⅔ cup lukewarm water
- Vegetable shortening/oil, for frying

Directions:

In a large bowl, mix together flour, baking powder, oil, and salt. Then pour in the warm water and stir everything until a soft dough forms.

On a surface dusted with flour, knead the dough for approximately five minutes, or until it is smooth and elastic.

For about fifteen minutes, cover the dough with a moist cloth and let it rest.

Once the dough has rested, divide it into 8 equal pieces and roll each into a round that is ¼ inch thick.

In a large skillet over medium-high heat, heat about ½ inch of oil or shortening until it reaches 350°F (175°C).

One round at a time, carefully drop it into the heated oil or shortening and cook for one to two minutes on each side, or until golden brown.

Before serving heated, remove from pan using tongs or a slotted spoon and place onto paper towels to drain excess grease. Have fun!

BILTONG

One of the greatest survival recipes is biltong, which has long been a favorite snack of hikers and outdoor enthusiasts and keeps them well-fed on adventures.

When combined, these components provide a really flavorful, high-protein meat jerky free of artificial preservatives and additives, making it perfect for extended trips into the outback. Furthermore, it's simple to understand why so many adventurers view biltong as a necessary component of their equipment given that it requires very little preparation before heading into the wilderness.

Ingredients:

- 2 lbs of lean beef (sirloin or round steak)
- 1/4 cup Worcestershire sauce
- 3 tablespoons white vinegar
- 2 tablespoons kosher salt
- 2 teaspoons ground black pepper
- 2 teaspoons sugar
- 1 teaspoon garlic powder
- 1 teaspoon allspice

Directions:

The meat must first be prepared by removing any visible fat and slicing it into pieces no thicker than an inch.

Next, in a shallow bowl, whisk together the Worcestershire sauce, white vinegar, sugar, salt, pepper, garlic powder, and allspice until well combined.

After putting the meat pieces in the marinate mixture, refrigerate for two hours.

Take the meat out of the marinate mixture and blot dry with paper towels after two hours.

Every piece of meat should be hung in an area with good ventilation on hooks or drying racks.

When the required amount of hardness is reached, let it dry for at least three days. Then, slice it across the grain into thin strips and enjoy!

CORNMEAL MUSH

One of the most traditional, straightforward, and satisfying survival meals is cornmeal mush. It is simple to prepare with a few ingredients and is high in vitamins and carbs.

Before serving, drizzle it with butter for even more flavor. It makes sense that one of the greatest survival meals you can cook is cornmeal mush, given how adaptable it is as a meal alternative during times of scarcity!

Ingredients:

- 1 cup cornmeal
- 1 tsp salt
- 4 cups boiling water

Directions:

When you've assembled all of your ingredients, it's time to cook! First, fill a medium saucepan with 4 cups water and heat it to a boil over high heat.

When the water is boiling quickly, turn down the heat to low and slowly stir in your cup of roughly ground cornmeal, whisking or using a wooden spoon to stir constantly.

As you add the grain, make sure to stir often to prevent lumps or clumps from forming in the mixture. This should not take more than five minutes to complete.

Reduce the heat even lower so that the water simmers gently after all of the cornmeal has been added and mixed (you don't want it bubbling too violently).

After adding your teaspoon of salt, cook over low heat, stirring occasionally, for about 25 minutes, or until the majority of the liquid has been absorbed.

Take off the stove and allow it to cool for roughly ten minutes before reheating and serving in separate bowls.

BANNOCK

A quick and tasty survival dish that you can prepare on the stove or over a fire is bannock. It's a simple and healthful method to eat when you're stranded in the wilderness or don't have access to other meals.

Because bannock may be served with almost any kind of topping—from meat to fruit preserves—any wet or dry item you may encounter along the wilderness trail, it's one of the greatest survival recipes to know.

Ingredients:

- Flour
- Baking powder
- Salt
- Bacon grease or lard

Directions:

To make the batter, first combine the dry ingredients (flour, baking powder, and salt) in a bowl and stir until blended well.

Next, include your fat (lard, butter, or oil) into the mixture and blend it with a fork or your hands until it resembles coarse crumbs.

To form a workable dough ball, add enough water or milk to the mixture.

If more liquid is required, add it until the mixture is workable; but, avoid adding too much liquid, since this will make the cooked product overly dense or heavy.

After combining all ingredients, place onto a surface dusted with flour and knead for approximately five minutes, or until the dough is smooth.

Next, form it into multiple smaller loaves or one huge round loaf, based on how many people you plan to feed with your homemade bannock!

Transfer to a hot skillet or baking sheet that hasn't been oiled, and bake for 20 to 30 minutes, or until the tops are golden brown. Serve warm with jam, butter, or honey.

RED BEANS AND RICE

Red beans and rice is a tasty and affordable recipe that's perfect for a survival kit. Only a few simple components are needed, the most of which are readily available in your kitchen pantry. The dish is quite substantial, particularly when accompanied by cornbread or jalapeño cornbread muffins. Additionally, because of the red beans, it is a great source of vitamins, minerals, and proteins.

It's also quite easy to cook; all you need is a pot and some spices to intensify the flavor. This implies that you can still prepare a delicious meal with very little equipment.

Ingredients:

- 1 lb dried red beans
- 1 large onion
- 4 cloves of garlic
- 2 tbsp olive oil or butter
- 2 tbsp ketchup
- 1 tsp cumin
- Salt and pepper to taste
- 2 cups of cooked rice

Directions:

Pour enough cold water into the pot to completely submerge the beans, then cover it and let them soak for the entire night.

Empty the saucepan of the soaking liquid the following day.

Add your butter or olive oil to a medium-sized pot and add your chopped onion. Heat the pot to medium. Add your cloves of garlic and sauté until tender.

Add your soaked beans and two quarts of fresh water after another minute or so. Bring everything to a boil and then turn down the heat to low-medium.

Add more water as necessary and simmer until the beans are soft but not mushy, stirring regularly, for about 1 hour and a half.

Add your ketchup, cumin, and seasonings last, and simmer for an additional five minutes before turning off the heat. Serve with cooked white rice on top!

PINOLE

Simple ingredients can be used to make pinole, a tasty and nourishing survival meal. Finely ground maize is what it is made of, and it is usually roasted in a pan with some oil. The finished product is flavorful and gives your body the much-needed carbohydrates and proteins it needs for energy. Because it doesn't need to be refrigerated, can be prepared anywhere, and has excellent nutritional qualities, pinole is one of the greatest survival recipes to know. Moreover, it's a fantastic way to finish off any leftover grains you might have lying around.

Ingredients

- 2 cups corn flour
- 7 tablespoons (3 ounces) piloncillo, pulverized into a fine powder
- 1 teaspoon powdered Mexican canela (cinnamon)

Directions:

Toast the corn flour in a skillet over medium heat for ten minutes, or until it becomes a deep tan that is a few shades darker than light brown sugar. After transferring to a big bowl, let it to cool fully. Whisk together the Mexican canela and the ground piloncillo until well combined.
And that's it!

DUMPLINGS

Dumplings are a great option if you're searching for a quick and delicious dinner that you can prepare with minimal ingredients! Not only are they easy and quick to make, but you can utilize up random materials you may have on hand because of their versatility.

Usually, you just need a little flour, grease (like butter or lard), and any kind of filling (such ground meat or cooked veggies). Dumplings are a great snack when cooked in a pan or boiled like Russian pelmeni or Chinese jiaozi and served with a side of sour cream.

Ingredients:

- 2 cups of all-purpose flour
- 2/3 cup of cold water

- 1 teaspoon of salt
- 1 teaspoon of baking powder

Directions:

In a medium bowl, begin by mixing together the flour, baking powder, and salt.

Add the cold water little by little until the mixture forms a dough. The dough should be soft and smooth after 5-7 minutes of kneading.

Roll out each of the eight equal portions of dough to form thin circles that are roughly 6 inches across. To keep it from sticking, add extra flour if necessary.

Put a tablespoon or more of your preferred filling in the middle of each round. You can use veggies like cabbage or carrots, cooked chicken or meat, or even dessert ingredients like apples or berries!

To create a semicircle, fold each circle in half over the filler. Using your fingers, firmly press the edges together to enclose the filling and stop any leaks while cooking.

Using tongs or a slotted spoon, carefully lower each dumpling into the boiling water one at a time (they should be done once they float up to the surface). Bring a pot of salted water to a boil over medium heat on your stovetop or over a campfire.

Warm up and serve!

CORN DODGERS

A traditional Southern delicacy that dates back to the early settlers, corn dodgers are still loved today for their flavorful, straightforward taste.

Because of its versatility and flexibility to be made with whatever ingredients you happen to have in your cupboard or out in the wild, this is one of the best survival recipes you'll ever come across. Corn dodgers are an excellent breakfast food on their own, especially when paired with a little honey. They also make a great side dish.

Ingredients:

- 2 cups finely ground cornmeal
- 1 teaspoon salt
- 1/2 cup lard or vegetable shortening

- 1/2 cup plus 2 tablespoons water
- Vegetable oil, for frying

Directions:

Stir the cornmeal and salt together in a medium-sized bowl. Using your hands, mix in the lard or vegetable shortening until it resembles coarse crumbs. Using your hands, slowly whisk in the water until the dough comes together into a ball. The dough shouldn't be overly dry or wet, just somewhat sticky.

On a surface dusted with flour, roll out the dough to a rectangle that is ¼ inch thick. Using a glass cup or biscuit cutter, cut into 3-inch circles. Make indentations with your fingers into each round of dough to create a pocket for the filling.

About one tablespoon of filling should be placed inside each ring of dough (see below). To prevent the filling from leaking out during cooking, fold up each side around the filling to form an oblong shape that resembles an envelope. Pinch any loose ends together.

Continue in this manner with each of the dough circles until they are all filled and firmly sealed.

In a large skillet over medium-high heat, heat enough vegetable oil to reach 375 degrees F (190 degrees C).

Gently drop two or three filled rounds at a time into the heated oil, and cook for approximately three minutes on each side, or until golden brown and crispy. Adjust the heat as needed to avoid burning or overbrowning.

Using a slotted spoon, remove from the oil and drain on paper towels. Serve warm, topped with your preferred honey butter or syrup!

PAN FRIED PORK CHOPS

This dish is easy to prepare, even during hard circumstances, since it only requires a few basic cupboard items.

Simply cook the pork chops in oil for about six minutes on each side over medium heat, or until browned and well done.

After that, take it out of the pan, season it with a little of your preferred herbs, and let it to cool.

This produces incredibly crunchy, golden-brown chops that can be eaten with other components or as a stand-alone meal.

Ingredients

- 2 lbs lean boneless pork chops, about 1.5 inches thick
- 1 teaspoon garlic powder
- ½ teaspoon salt
- ½ teaspoon pepper
- 2 tablespoons olive oil
- 2 tablespoons salted butter
- ½ teaspoon minced fresh thyme or rosemary

Instructions

After rinsing, wipe dry, and evenly season both sides of the pork chops with salt, pepper, and garlic powder.

In a big skillet, heat the oil over medium-high heat. Sear and brown the bottom side of the pork chops for one to two minutes.

Before turning again, cook for a further one to two minutes on the other side. Continue turning the chops until they are a deep golden brown and, after 8 to 10 minutes (cooking time will vary depending on thickness), an instant-read thermometer put into the thickest section registers 145°.

When the pork chops are cooked to your preferred doneness, add the butter and thyme and fry for a further two to three minutes, spooning the butter over the chops as they cook.

Transfer the pork chops to a serving plate and spoon over one last time the pan's butter-sauce.

After 5 minutes of resting on a foil-covered dish, serve the pork chops. (Letting them rest enables them to develop their full juicy and tender texture!)

POTATO CAKES

Because they are easy to make and don't need a lot of ingredients, potato cakes are a perennial favorite among survival recipes. This recipe works well as a midday snack or served as breakfast.

Ingredients

- 2 cups mashed potatoes
- 1 cup all-purpose flour
- 1 onion, diced
- 1 egg
- ½ teaspoon ground black pepper
- ½ teaspoon salt
- ½ cup vegetable oil, or as needed

Directions

In a bowl, thoroughly mix the mashed potatoes, flour, onion, egg, black pepper, and salt until completely incorporated, about the consistency of batter.

In a skillet, heat the vegetable oil over medium heat. Drop batter into the hot oil in 4-inch circles. Cook for 4 to 5 minutes on each side, or until golden brown; drain on paper towels. Proceed with any leftover batter.

OXTAIL POTJIE

The recipe for Oxtail Potjie is deceptively simple, yet it tastes very rich and hearty.

The original purpose of the traditional South African recipe was to provide visitors with an inventive means of cooking in the bush while using the fewest possible supplies.

The best part is that this dish is one of the best survival meals anyone can make because it just needs minimal attention while cooking in a big pot over an open flame.

Ingredients

- 1kg Oxtail joints
- 3 tbsp of olive oil
- 1/2 cup of flour (for dusting)
- 1 med onion chopped
- 3 cups of red wine
- 1 cup of water
- Freshly ground salt and pepper

- 1 tin of chopped tomatoes
- 3 tbsp of tomato paste
- 2 tbsp of crushed garlic
- 10 black peppercorns
- 2 bay leaves
- 3 sprigs of Rosemary
- 8-10 whole baby onions peeled
- 8 med-large carrots coarsley chopped into rounds
- 10 -12 baby potatoes unpeeled and halved
- 2 cups of green beans with ends removed
- 25g packet of oxtail soup powder
- Freshly ground salt and pepper

Directions

As you get ready to light your potjie, make sure you have a enough supply of hot coals by using wood, briquettes, or charcoal.

After cleaning the oxtail joints, give them a quick pat with a paper towel and a dusting of flour. Add the olive oil, chopped onion, garlic, and oxtail to the pot and heat it over a fire.

Be careful not to burn the chunks of oxtail as you gently brown them.

Add the red wine, water, tomatoes, tomato paste, peppercorns, bay leaves, and rosemary after the meat has been wrapped.

Add a pinch of finely ground salt for seasoning.

Put the lid on and simmer for around two and a half hours over a moderate to low heat. Periodically check to see if more water needs to be added.

The meat should not be drowned; it should merely be covered in liquid.

Add the baby onions, diced carrots, and halved baby potatoes at this point, then cook for an additional hour. Add the oxtail soup to the saucepan with the green beans, mix it with the water, and simmer for about an additional hour or so, or until the meat starts to fall off the bone.

If necessary, add freshly ground pepper and salt for seasoning.

Accompany with handmade pot bread and rice.

FISH ON A STICK

Just a few basic supplies, including fish filets or your favorite seafood, skewers, and your favorite marinade, are needed to make this beloved outdoor delight.

Ingredients

- 18 ounces cod or other lean white fish, cut into 12 equal pieces
- ¼ teaspoon salt
- ¼ teaspoon freshly ground black pepper
- ½ cup raw pumpkinseeds
- 2 teaspoons grated lemon rind
- 1 tablespoon chopped fresh parsley
- 1 tablespoon olive oil
- ½ cup low-fat tzatziki sauce

Directions

Warm up the broiler. Put a baking sheet with a rim inside of a metal rack.

One fish piece should be threaded onto each of the twelve (6-inch) skewers (you can trim the sharp ends of the skewers in advance if you'd like); season the fish with salt and pepper.

In a food processor, add the pumpkinseeds and pulse until finely crushed. Move to a shallow plate and mix in the parsley and lemon rind. In a shallow dish, add olive oil.

Coat the fish evenly by dipping it first in olive oil and then in the seed mixture. Put on the pan's rack. Continue with the remaining skewers. Apply frying spray to the fish on both sides. Broil for two minutes on each side, or until the fish is firm and the pumpkinseeds start to brown. Add tzatziki sauce and serve.

Vetkoek

A popular meal from South Africa that is enjoyed by people worldwide is called vetkoek. Because you can make it with very basic ingredients—most of which you probably already have in your pantry or refrigerator—it's an excellent survival meal.

Ingredients

- 2 cups lukewarm water
- ¼ cup white sugar
- 1 (.25 ounce) package active dry yeast
- 7 cups all-purpose flour
- 2 teaspoons salt
- 3 cups oil for frying

Directions

In small basin, mix yeast, sugar, and lukewarm water. Allow it stand for about five minutes, or until yeast begins to soften and bubble.

In a large basin, sift together flour and salt.

Mix the yeast mixture with the flour mixture and knead for 5 to 7 minutes, or until the dough is elastic and smooth. Allow dough to rise in basin covered with a clean cloth until doubled in volume, about 45 minutes.

About the size of a tennis ball, pinch out a piece of dough and roll it until smooth. Roll dough ball into palm-size ball and place on a floured surface. Continue with the leftover dough.

Preheat oil in a big saucepan or deep fryer to 350 degrees Fahrenheit (175 degrees Celsius).

Fry two to three pieces at a time in heated oil for about three minutes on each side, or until golden brown. blot with paper towels.

RUSKS

Hardtack, sometimes referred to as ship's biscuit or rusks, is a classic seafaring staple. They are the ideal meal for any sailor on a lengthy cruise, or even for those wishing to plan for the future, as they are very nourishing and easy to prepare.

The best part is that there is little no preparation needed for the main components required to produce rusks: All you need is baking soda, salt, water, and flour. These ingredients are combined in a bowl, formed into a dough, then baked until crisp and dry.

Ingredients

- lbs self-raising flour

- lb pure butter, firm
- cups sugar
- teaspoons salt
- large eggs, beaten
- tablespoon vanilla
- cups buttermilk, to mix in (about)

Directions

350 degrees Fahrenheit in the oven.

Butter two large, level cookie tins.

Grate the butter into the flour in a large bowl using the coarse side of a grater.

After that, work the mixture in with your hands until it resembles breadcrumbs.

Add the sugar and salt.

Mix the vanilla into the beaten eggs and then, very coarsely, stir into the flour mixture.

Add enough buttermilk to the mixture to get the soft, somewhat sticky, but not moist, consistency of scone dough. It should yield about 6 cups, but you may need to adjust for that.

Prepare a bowl of relatively warm water and submerge your hands into it. Now, gently form the dough into large balls.

The balls ought to be roughly one-third the size of tennis balls (sorry, tough if you can't show it!) Place these in the oiled tins, touching each other. Packing the dough too closely to the tin's edges can cause it to rise significantly.

A few factors, such the size of the balls, affect how long they bake; in general, 45 to 60 minutes The tops of the rusks should be golden-brown and well-risen. Take a glance, but keep it from burning. Use a skewer to test.

BUFFALO JERKY

Buffalo jerky is a great recipe for a survival kit since it's quick to prepare, packed with nutrition, and keeps well.

Ingredients:

- 1.5-2 pounds top round beef, London broil
- 1 cup buffalo sauce
- kosher salt

Direction:

Slice the meat more easily by freezing it for approximately one hour.

Take out of the freezer and slice into long, thin pieces.

Toss the beef to ensure that every piece is covered in buffalo sauce after placing it in a big basin, adding salt to taste and buffalo sauce. Allow to marinate for at least 6 hours, or overnight.

Put each beef piece on the dehydrator tray (*see note below for an alternate oven method) and let it dry at the temperature recommended by your machine, which is typically 160 degrees. After 3–4 hours, check the crispness of the beef. This is the moment where it might be ready, depending on how thin you cut it. Thick, mushy portions should be kept in the dehydrator until they are completely dry.

For three to four weeks, store in an airtight jar.

Notes As an alternative, roast the marinated beef for four to six hours at 175 degrees on a cooling rack set over a baking sheet (or until fully dried).

PIONEER BREAD

Pioneer bread makes a wonderful and nourishing option for a survival recipe. It requires very few ingredients and preparation, and it keeps for weeks without needing to be refrigerated.

INGREDIENTS

- 8 tablespoons cornmeal
- Pinch salt
- Pinch baking soda
- 1 teaspoon sugar
- Scalding hot water
- 1 quart milk
- 1 1/2 teaspoons salt, divided
- 4 cups all-purpose flour

- 3 to 4 tablespoons shortening, butter or lard
- Additional flour, if needed
- Butter for tops of loaves

Direction:

The evening before you want to bake:

Stock a pint jar with cornmeal. Add a few dashes of soda and salt. Put some sugar in.

Pour in enough boiling water to cover the jar. Mix thoroughly. Place a lid on it and leave it in a warm location for the entire night.

Mix milk, cornmeal mixture, flour, and 1/2 teaspoon salt together in the morning. To make a firm batter, stir thoroughly.

Place the entire bowl in a jar with warm water, cover it, and keep it somewhere warm. It's probably going to be necessary for you to periodically remove part of the water and replace it with warmer water. Make an effort to maintain the temperature as steady as you can. It will climb in three to four hours. Re-stir if it is rising very slowly, and then return the bowl to the warm water.

After the dough has risen, mix one teaspoon of salt, enough shortening, and flour to form a stiff dough. Knead for ten minutes.

Turn the oven on to 375°F.

Cut dough into three loaves. After thoroughly greasing the loaf pans' edges with butter, put the loaves inside and let them rise for 45 to 1 hour, or until they have doubled in size.

Bake for thirty to forty minutes. As soon as the bread is done baking, remove it from the oven and generously butter the crust.

JOHNNY CAKES

Johnny Cakes also have the benefit of not requiring any specialized cooking equipment—you can bake them in a pan over an open flame without even using an oven!

This delectable delicacy, a classic New England favorite, only requires flour, sugar, and either milk or water to prepare. It's also incredibly flexible. The majority of ingredients are easily located in every kitchen cupboard, and they require little preparation time.

Ingredients:

- 1 cup of white cornmeal
- 1 teaspoon of baking powder
- 3 tablespoons of sugar
- ½ teaspoon of salt
- 2 tablespoons of melted butter

Directions:

In a bowl, combine first all of the dry ingredients.

After that, melt the butter and gradually stir it into the dry ingredients until all of the ingredients are combined. Small amounts of milk or water can be added as needed to achieve a thick yet pourable consistency for the batter.

Spoonfuls of batter are added to a skillet that has been heated with oil or bacon grease over medium-high heat.

Fry them till golden brown on both sides, then turn them over and continue cooking the other side. Serve warm with honey or butter.

VENISON STEW

A tasty and very nutritious survival meal, venison stew is full with critical vitamins and minerals for human health. Therefore, one of the greatest recipes you can create is venison stew in case you find yourself in a position where you have to rely on food sources other than grocery shops or restaurants. Simmering venison meat with basic ingredients such as onions, potatoes, carrots, garlic, red wine, or broth can be a simple and quick preparation method. Feel free to experiment with other combinations!

By combining all the ingredients in perfect harmony, the stew's richness of tastes softens the gamey flavor of the meat.

Since venison stew may have just about anything added to it, it's undoubtedly one of the best foods to eat during hard times. It is therefore filling and nourishing.

Ingredients:

- 2 pounds of cubed venison
- 2 tablespoons of olive oil (or other cooking oil)
- 1 large onion (diced)
- 2 cloves of garlic (minced)
- 1 teaspoon of dried thyme leaves
- 2 bay leaves
- 1 tablespoon of tomato paste
- 4 cups of beef broth or stock
- 4 carrots (sliced into half-inch pieces)
- 2 medium potatoes (cut into cubes)
- 1 cup of frozen green peas
- Salt and pepper to taste

Directions:

To begin, place a big saucepan over medium-high heat with the oil. Cook the chopped onion until it becomes tender.

After that, toss in the cubed venison after adding the garlic and cooking for an additional minute. Cook until the outside of the venison is just beginning to brown.

Next add the tomato paste, bay leaves, thyme leaves, and beef broth or stock to the pot. Bring to a boil and then turn the heat down to low. Once the beef is cooked, simmer it for one and a half hours with a lid on.

Put potatoes and carrots in the pot. Reapply the lid, then boil the vegetables for a further forty minutes, or until they are soft.

Add frozen green peas and, if preferred, season with salt and pepper.

Cook for an additional ten minutes and serve.

DRIED FRUIT

Because dried fruit is inexpensive, wholesome, and has a lengthy shelf life, it's an excellent survival ingredient to know. It's simple to make your own dried fruit; you may use the sun, an air fryer, a dehydrator, or even an oven!

Ingredients:

- Fruit of your choice (apples, apricots, bananas, berries, etc.)
- Sugar (optional)
- Citric acid (optional)
- Salt (optional)

Directions:

You can dry whatever kind of fruit you want to. After giving it a thorough wash, slice or cube it. Before drying, it's preferable to soak some fruits, such apples, in water containing a tablespoon of lemon juice or citric acid for around fifteen minutes. This will keep them from browning and help maintain their color. Before drying the fruit, you can optionally sprinkle it with salt or sugar.

Adjust the oven temperature to 150–175 degrees Fahrenheit based on the kind of fruit you are drying and its moisture content. The fruit will dry more slowly at lower temperatures, but it will also help avoid scorching or over-drying.

Arrange the fruit, either sliced or cubed, on silicone mats or parchment paper-lined baking sheets. To allow air to flow freely around the pieces while they dry out in the oven, make sure there is plenty of room between them.

Depending on how thickly you slice or chop your pieces, baking time may vary from 2 to 10 hours. Thicker pieces will require more time! Throughout the baking process, check frequently to ensure sure they aren't burning or overdrying.

After baking, let the dried fruit pieces cool fully before keeping them for up to six months at room temperature in an airtight container! If preferred, dried fruits can also be frozen for up to a year.

PEASANT BREAD

One of the simplest and best survival recipes is Peasant Bread. It only needs five basic ingredients, and you can make it without an oven!

In addition to tasting delicious, this bread stores extremely well and is nearly impossible to rot as long as it is kept out of the wet. This makes it an excellent option for use in emergency situations. On surfaces, it can last up to a week; if you refrigerate or freeze it, it can last for months.

Ingredients:

- All-purpose flour
- Baking powder
- Salt
- Butter or oil
- Warm water
- Honey

Directions:

Set the oven's temperature to 350 degrees Fahrenheit, or 180 degrees Celsius. This ought to take ten minutes or so.

Combine 2 cups all-purpose flour, 1 teaspoon baking powder, and ½ teaspoon salt in a medium-sized bowl. You can change the salt quantity to suit your preferred level of salinity in your bread.

Mix with 2 tablespoons oil or butter until the mixture resembles coarse crumbs. Add the sugar or honey now if using either. For a mild sweetness, one tablespoon should do, but you may certainly alter to suit your tastes.

Pour in 1 cup of warm water slowly, stirring with a wooden spoon until the mixture forms into a ball of dough that is somewhat sticky but not moist.

Shape the dough into a loaf that is about 8 inches long, 4 inches wide, and 2 inches high. Place the loaf onto an ungreased cookie sheet that has parchment paper laid on it after kneading it for 3 minutes on a lightly floured surface until it is smooth and elastic.

Bake for 25 minutes at 350°F (180°C) or until golden brown; remove from oven and let cool completely before slicing and serving.

HASTY PUDDING

Because it only requires a few ingredients and can be made quickly—hence the name—hasty pudding is among the greatest survival recipes to know.

Ingredients

- 2 cups whole milk
- 1 cup heavy cream
- 1/2 cup molasses
- 1/4 cup dark brown sugar
- 1/3 cup cornmeal
- 1 teaspoon ground ginger
- 1 teaspoon cinnamon
- 1/4 teaspoon salt
- 1 pint vanilla ice cream

Directions

Preheat oven to 350°F. Heat the milk, cream, molasses, and brown sugar in a medium-sized, heavy stainless-steel saucepan over a moderate heat, stirring from time to time, until the mixture almost simmers.

Mix the cornmeal, ginger, cinnamon, and salt in a medium-sized bowl. While whisking, add to the milk mixture. Whisk and bring to a gentle simmer. Transfer to an 8 × 8-inch baking pan. The batter will have a shallow consistency.

For twenty minutes, bake the pudding in the center of the oven. Take out of the oven and thoroughly stir. Put the pudding back in the oven and let it cook for a further 20 minutes. Even though it's still quite jiggly, the pudding will solidify as it cools. After 20 minutes of cooling on a rack, serve warm. Alternatively, let the pudding cool fully and then reheat it for approximately five minutes at 350°F right before serving. Present the pudding with the ice cream on top.

OLD FASHIONED STACK CAKE

Making this tasty delicacy only takes a few minutes and calls for simple items that will nourish both body and soul.

Ingredients:

- 2 cups all-purpose flour
- 1 teaspoon baking powder

- 1/2 teaspoon baking soda
- 1/4 teaspoon ground cinnamon
- 1/4 teaspoon salt
- 3/4 cup butter, softened
- 2 cups firmly packed light brown sugar
- 3 eggs
- 3/4 cup buttermilk or regular milk

Directions:

Grease two 8-inch round cake pans lightly and preheat the oven to 350°F.

Mix the flour, baking soda, baking powder, cinnamon, and salt in a medium-sized bowl and set it aside.

Beat butter and sugar together in a big bowl until light and fluffy. One egg at a time, add to mixture and mix on low speed until well incorporated.

After adding buttermilk (or plain milk) and mixing until incorporated, add half of the dry ingredients to the wet components and stir until barely blended. Mix in the remaining dry ingredients until just combined, being careful not to overmix.

The mixture will be thick, so divide it equally among the prepared pans and use a spoon or spatula to spread it out into an even layer. After preheating the oven, put the pans in and bake for 25 minutes, or until the tops are golden brown (if needed, test with a toothpick). Allow to cool fully before putting the stack cake together.

Melt butter, brown sugar, and cinnamon in a small bowl for the filling; put aside.

Once the cakes have cooled down fully, start assembling your stack cake by putting a quarter of the filling on one side of each layer of cooled cake (the layers should face upwards to facilitate the spreading process).

Stack the first layer, filled side down, on the serving dish, followed by the second layer, filling side up. Over the second layer, spread the remaining filling and, if desired, top with walnuts. If storing overnight, serve right away or securely cover with plastic wrap before serving.

Emergency Cooking

Imagine having to cook meals for your family without electricity, outside, and from scratch. Is imagining it difficult? What if the scenario included the stress of a natural disaster? Without the modern conveniences you take for granted now, could you prepare meals with the knowledge, abilities, and tools you currently possess? Basic foods like wheat and beans, which require more time and effort to cook, are among the foods suitable for long-term storage. Other options include freeze-dried meals that require only water. Similar to this, emergency cooking can be as simple as hand-grinding wheat for bread baking in a Dutch oven or as complex as boiling water over an open flame to rehydrate a freeze-dried meal. The kinds of food you have on hand, your level of expertise, and the tools you have will determine how you cook in an emergency. But don't be alarmed. For now, the following fundamentals:

KNOWLEDGE

If you don't know how to use them, no amount of equipment or food storage will be of any use to you. Which foods you store, along with the skills and tools you'll need, will depend on what and how you prepare. Thus, it's critical to first understand who you are and your family. After determining which kinds of food storage you'll employ and store, it's critical to gain as much knowledge as you can about them. Reading product manuals, recipes, and other books and articles about food storage and emergency cookery are a few examples of this. You'll be more prepared the more you learn.

SKILLS.

Although knowledge is precious, you should put it into practice. You may get comfortable with emergency cooking before a crisis arises by learning how to set up your food storage supplies, experimenting with recipes, and using your equipment. It will also assist you in optimizing your plan for food storage.

EQUIPMENT.

Again, the kinds of cooking utensils you require will vary based on the kinds of food you store and how you prepare. You will require a grain mill, baking supplies, and other items if you intend to limit your food store to staples like wheat, beans, oil, etc. You may not need anything more than a camp stove and a pot to boil water if the majority of the food in your store is just-add-water items. It's critical to investigate your alternatives by becoming knowledgeable about and experimenting with a range of emergency cooking techniques for various storable food kinds.

Here are a few substitute cooking techniques for natural catastrophes and emergencies:

Camping stove

An excellent emergency and disaster resource is a camping stove setup powered by propane or butane. It's reasonably priced, simple to use, quick to set up, and effective. You might get a large camping stove, a simple hikers burner (complete with a bear bowl or cook pot), or even a Jetboil. It all relies on your choices and financial situation.

Remember to purchase enough of butane or propane. And cook as much as you can with this system to really test your equipment.

Use a camping stove setup that has enough of airflow, ideally outside or next to an open window. However, proceed with caution.

Convert a Can into a Campfire.

A camping stove from the store is a great invention. Most are effective and simple to use. But you must know how to construct a cook stove and locate substitute fuels for it in case you don't have access to one or run out of fuel. Cans of food and drink are the basic material for many stoves that burn wood and other dry fuels as well as liquid fuels. All you need to do is cut the metal correctly, fill it with gasoline, and fire it. Sticks and other plant fuels can be burned in homemade "rocket stoves." Fuels such as alcohol and cardboard soaked in wax can also be burned by them. Just remember that adding gasoline or other explosive fuels to a stove like this might have disastrous consequences.

Fire/BBQ/Rocket Stove.

Since they all require the outdoors and some kind of wood or charcoal, we combined them all into one category. It will be quite effective to use a BBQ or fire pit that you already have at your house or flat. However, this is only useful if you can walk outside and utilize it. It's likely that you won't be able to utilize this if it's storming heavily outside.

A rocket stove is quite simple to construct and operates incredibly well.

Make sure you have an ample supply of the fuel sources, such wood or charcoal, on hand.

Sun.

Many people disapprove of solar cooking, however it functions similarly to a slow cooker. With a solar oven, you can bake, cook, and even sterilize water. You can build your own solar oven; Google searches yield a plethora of instructions.

This must be done outside in the full sun, much like cooking over a fire or with charcoal. With clouds, cooking is possible, but it will take twice as long.

Pans and Griddlers.

With the use of skillets, frying pans, griddles, and homemade frying containers, you may prepare delicious dishes that frequently resemble comfort food from home. A frying pan or griddle suspended over a small fire works well when supported by bricks, stones, cinderblocks, or other fire-resistant materials. If these supports are just 6 inches tall, you can still cook with them, but supports that are one foot tall will function considerably better.

While they aren't as reliable, the following additional sources can be useful in an emergency:

Candles: Boiling water with candles takes around an hour and results in uncooked rice, although it is technically cooked. It is a time-consuming and resource-intensive method that you may consider if you have exhausted all other options. suitable for indoor usage.

Buddy burner: You can use this indoors and it's really simple to create your own. It is significantly more efficient to do this than to use candles.

The alcohol stove is reasonably effective. They do function effectively, but they require a lot of liquid. Making your own alcohol stove is fairly simple.

Cooking with Foil: All you need is some food and a cooking fire to transform a basic sheet of foil into an oven, broiler, frying pan, and many more cooking utensils. Using foil pouches to cook delicious meals over coals is a tremendous improvement over baking food directly in the ashes or embers of a fire. Here are some pointers:

In order to avoid suffering or having to work more than necessary during an emergency or disaster, keep in mind that if you want to survive, you must be ready with the right tools and resources.

Thoughts:

When a disaster strikes, having the ability to prepare delectable meals using only the supplies you have on hand is a very useful talent. These dishes may be utilized in everyday life, even if disaster never hits. Just think of the happiness you'll feel after creating them!

Get your practice going now!

BOOK 8: FIRE-MAKING MASTERY

For everyone who spends time outside, fire building is a crucial survival skill. The ability to start and keep a fire is essential whether you're going camping, trekking, or just having a backyard BBQ. There are a few different ways to start a fire, and the one you use will rely on the circumstances.

The teepee method is a popular fire construction technique in which larger sticks are leaned against little kindling in the middle of the fire pit to create the form of a teepee. This is an excellent way to light a fire quickly and effectively. It also performs well in windy environments. An alternative approach is the "log cabin method," which involves piling logs into a square shape and placing kindling in the middle. Building using this approach might be difficult in rainy weather, but it is perfect for longer-lasting flames.

It is important that you understand how to properly maintain your fire, regardless of the approach you use. This entails maintaining the fire contained within your fire pit, controlling the airflow, and adding fuel as necessary. Anyone can learn to make and sustain a fire in any situation with the correct methods and some practice.

FUNDAMENTALS OF FIRE BUILDING

CHOOSING A SITE

Choosing the ideal location is the first step in starting a fire. The area has to have good ventilation and be kept clear of any combustible objects, such as branches, dry leaves, or grass. Additionally, picking a location away from the wind is crucial since wind can make it harder to start and keep a fire going.

FIRE MATERIAL SELECTION.

It's time to start gathering firewood after you've chosen a location. To ignite a fire, three essential ingredients are required: fuel, kindling, and tinder. Tinder, which might be little twigs, leaves, or dried grass, is what will start a fire. The next level up is called kindling, and it usually consists of little sticks or branches. The biggest material, fuel, such logs or huge branches, is what will keep the fire going.

ARRANGING YOUR MATERIALS.

It is crucial to correctly arrange your items once you have gathered them. The tinder should first be placed in the middle of the fire pit. After that, arrange the kindling in the shape of a teepee or log hut around the tinder. Lastly, place the fuel over the kindling.

These basic procedures may be used to start a fire in a variety of situations. It's crucial to keep in mind that you should never leave a fire unattended and that you should completely put out a fire before leaving the area.

FIRE BUILDING TECHNIQUES

For anybody spending any time outside, knowing how to build and manage a fire is a must. A successful fire may be started and maintained by knowing the various fire layouts and when to utilize them.

Teepee Fire Lay.

One time-tested and highly successful fire-building technique is the teepee fire lay. It is particularly helpful in windy weather and is perfect for rapidly lighting a fire. A bundle of tiny, dry kindling should be placed in the center of the fire pit before beginning to create a Teepee fire lay. Next, create a cone shape by bending bigger sticks or logs around the kindling. To let air in and ignite the kindling, leave a gap on one side of the teepee. Gradually fill the Teepee with bigger wood as the fire becomes bigger.

Log Cabin Fire Lay.

Another common technique for starting a fire is the log cabin fire lay. It is perfect for cooking since it gives pots and pans a sturdy base. The first step in creating a log cabin fire lay is to align two enormous logs parallel to one another, leaving a tiny space between them. In order to create a square, add two more logs perpendicular to the previous two. To create a square stack of logs, keep alternating the logs in this manner. Light the tiny kindling that has been placed in the center. Gradually add larger logs to the stack as the fire becomes bigger.

Lean-To Fire Lay.

Building a fire is made easy and efficient using the Lean-To fire setup. It is fast to construct and perfect for windy days. First, locate a large, strong wood or rock to rest against in order to construct a rest-To fire setup. Next, tilt a small piece of kindling toward the ground and press it against the log or rock at a 45-degree angle. As the fire gets bigger, add bigger sticks and logs after lighting the kindling.

Platform (Upside Down) Fire Lay.

There is a special way to start a fire that is called the Platform fire lay. It can assist in keeping the fire off the ground and is perfect for damp or snowy weather. First, arrange two huge logs parallel to one another with a little space between them to start building a platform fire lay. In order to create a square, add two more logs perpendicular to the previous two. To create a square stack of logs, keep alternating the logs in this manner. Turn the pile over such that the logs are stacked on top of one another. Light the tiny kindling that has been placed in the center. Gradually add larger logs to the stack as the fire becomes bigger.

Anyone can start and sustain a successful fire in a variety of settings by being aware of and using these diverse fire lay techniques.

IGNITION METHODS

There are several approaches to choosing from when it comes to igniting a fire. Certain approaches could work better in a certain circumstance than others. We'll examine and weigh the benefits and drawbacks of some of the most popular igniting techniques in this section.

Lighters and matches

Lighters and matches are the most often used and practical tools for lighting a fire. They fit neatly into a pocket or bag and are simple to use. Matches are particularly helpful in damp weather since they may be wax-proofed or kept in a waterproof container.

Friction-Based Methods

In friction-based approaches, tinder is ignited by rubbing two things together to generate heat. The bow drill and hand drill are the most often used friction-based techniques. In dry circumstances, these techniques may be highly effective, but they do take expertise and experience.

Steel and Flint

A classic fire-starting technique is flint and steel. To ignite tinder, a piece of steel is struck on a piece of flint to produce sparks. In dry conditions, this approach may be highly effective, but it does need some skill and experience.

Solar Ignition

Utilizing the sun's rays to direct light onto tinder and ignite it is known as solar ignition. Although this technique works best in dry weather, it does require clear sky and bright sunshine.

Overall, there are benefits and drawbacks to each ignition method, and the appropriate approach may vary according on the circumstances. It is important to possess a diverse range of techniques and to hone their application before to venturing into the wilderness.

FIRE MAINTENANCE

Once a fire is started, it has to be properly maintained in order for it to burn safely and effectively. Maintaining a fire entails regulating its intensity, adding fuel effectively, and carefully putting it out.

Controlling Fire Intensity

It is crucial to manage a fire's intensity for both efficiency and safety. Overly hot fires can be hazardous, and under-intense fires can not produce enough heat or light. A fire's intensity may be managed by varying the quantity of oxygen accessible to it. The intensity of the flames can be lowered by partially enclosing the fire with a piece of wood or other material, so reducing the quantity of oxygen reaching the fire.

Adding Fuel Efficiently

In order to maintain a fire going, fuel must be added, but doing so must be done safely and effectively. A useful method for adding fuel is to use tiny, dry branches and twigs. These will ignite rapidly and offer a consistent supply of fuel for the fire. Once the fire is going, you may add larger chunks of wood. It is crucial to refrain from dousing the fire with too much fuel at once, since this might put out the flames.

How to Put Out a Fire Safely

It's crucial to put out a fire carefully to stop it from spreading and to prevent injuries. Applying water to a fire is one method of putting it out. It's critical to use enough water to completely put out the flames and prevent any embers from burning. Smothering a fire with sand or soil is another method of putting out a fire. For tiny flames, this approach works well, but it might not work well for larger ones.

All things considered, maintaining and starting a fire under various circumstances requires careful fire maintenance. A fire may be a helpful tool for cooking, giving warmth, and calling for assistance in an emergency by managing the fire's intensity, adding fuel effectively, and extinguishing it carefully.

Adapting to Weather Conditions

The weather must be taken into account before starting a fire. Different approaches to starting and sustaining a fire are needed depending on the weather. The following advice can help you adjust to varying weather conditions:

Wet Weather Fire Building

While it might be difficult, building a fire in the rain is not impossible. Locating fuel and dry kindling is crucial. Seek for dry wood in secluded spots or behind trees. To expose the dry core layer of wood, you can also use a knife to shave off the outer layer of wet wood. It's also crucial to utilize a fire starting made for damp environments, such a magnesium fire starter or waterproof matches.

Wind-Affected Fire Building

It might be challenging to start and keep a fire in the wind. Constructing a windbreak is one method of dealing with wind. You may accomplish this by arranging logs or pebbles around the fire. Constructing a wind-sheltered fire pit is an additional choice. Using fuel and dry kindling is also essential since wind may put out a fire rapidly.

Cold Weather Fire Building

A different strategy is needed to start a fire in cold weather than it is in warm weather. Locating fuel and dry kindling is one of the main obstacles. Seek for dry wood in secluded spots or behind trees. To expose the dry core layer of wood, you can also use a knife to shave off the outer layer of wet wood. It's also crucial to utilize a fire starter made for cold climates, such a lighter made specifically for usage in the winter.

In conclusion, depending on the weather, different methods are needed to start and keep a fire. Even in inclement weather, you can create a fire that will keep you warm and give light by using these suggestions.

BOOK 9: PREPPER'S NATURAL MEDICINE

We've all experienced stuffy noses, itchy throats, and uneasy stomachs. Most individuals have go-to at-home cures when they're feeling under the weather, just like us. Most focus on herbs, such as how they work better than over-the-counter remedies to soothe upset stomachs. Throughout history, people have utilized herbs for their therapeutic benefits for more than 5,000 years. While certain ailments necessitate a trip to the doctor and prescription drugs, you may sometimes get by with using plants' healing power at home, without even leaving the comforts of your sofa.

Types of Herbs to Utilize

Just as there are several methods to go from place to place, there are also numerous ways to incorporate herbs into our systems. However, you should utilize the form with the fastest response time in an emergency.

It is not advisable to prepare tea for emergency consumption. It is recommended that you include those herbs into your system as soon as feasible; they will be in the form of tinctures. Consider it like delivering the herbs by jet plane as opposed to by mule.

Powder: Dried or powdered herbs work well as topical treatments or as an ingredient in tinctures or other preparations where the alcohol would not be helpful. Because tinctures containing alcohol sting open wounds, the powdered version is preferable in this situation.

Tincture: Tinctures work well for small-dosage, portable usage. Tinctures have a 20-year shelf life and are absorbed by the body more quickly.

Upset Stomach

Ginger

Ginger may do wonders for a stomachache. Important enzymes and substances known as gingerols are found in the aromatic root. These substances start to act when ingested, reducing nausea and successfully eliminating gas from the body. When combined, these two processes facilitate digestion and hasten your recovery. Chew on a 1-inch piece of ginger to experience these advantages, or slice up and simmer in hot water to make a tea.

Fortunately, growing ginger at home is really easy! Make two-inch cuts first. To create a callous, let the cuttings out overnight. After drying, plant your cuttings in soil, give them plenty of water, and sunlight. You ought to see sprouts in two weeks!

Give your ginger plant a place on a windowsill after it has established to guarantee plenty of sunshine. Compost is also an excellent addition for ginger plants, as they like soil that is high in organic matter. Lastly, watch careful that your ginger plant doesn't become dry. Early development requires soil that is both moist and well-draining.

Bay Leaf

With a long history, bay leaves were revered as a regal plant in ancient Greece and Rome. Laurel crowns were also made using bay leaves. These days, dried bay leaf is frequently used as a flavor to a range of soups and stews. In addition to adding flavor to food, bay leaf has qualities that help soothe upset stomachs. These qualities originate from enzymes linked to the alleviation of gas and cramps in the abdomen. Similar to ginger, bay leaves make a useful tea when steeped in hot water. Your kitchen will smell great as well as have a satisfying feeling in your tummy!

The bay tree produces the bay leaf. The bay tree is native to Asia and requires lots of space to develop. The bay tree also needs lots of water, complete sun exposure, and organic matter-rich soil. Winter is coming, therefore it's also crucial to remember that bay trees are robust plants that can withstand freezing temperatures. Since the wood of the bay tree is not well-known for its strength, it must be protected from windy environments.

Congestion

Peppermint

Although it's a common ingredient to holiday confections, peppermint is a really beneficial plant for congestion sufferers. Menthol, which is found in peppermint, is well recognized for significantly enhancing nasal airflow. Furthermore, peppermint is an excellent remedy for overactive sinuses due to its antibacterial and anti-inflammatory qualities. Boiling water may be steeped peppermint leaves to produce aromatic vapor. Deep breaths and covering your face with the steam allow the peppermint to do its thing.

It is recommended to plant peppermint indoors in a container. This is because, when left outside, it has invasive qualities. A spot with full sun is ideal for your peppermint. Never allow the soil dry up on your peppermint plant; make sure it gets enough of water.

Eucalyptus

Eucalyptus is much more than just a lovely ornament. The eucalyptus tree's oils are well renowned for their ability to effectively combat congestion. The oil that is generated by eucalyptus leaves contains traces of menthol and other chemicals, just like peppermint. These substances are necessary for unclogging sinuses and facilitating better breathing. Gather whole eucalyptus branches into a bundle, knot the ends together, and hang it in your shower to get the full effects of the plant. The water's steam will stimulate the leaves to exude their oils.

Similar to the bay tree, eucalyptus is not a plant that does well inside. The eucalyptus tree may reach heights of 20 feet and widths of the same. As a result, you should make sure your tree has enough room to grow, ideally in a sunny spot. The eucalyptus tree prefers well-drained, somewhat acidic soil after a spot is chosen.

BURNS

Aloe

The sun's beams have the same intensity whether you're lounging on the beach or working in your yard. Everyone is aware of how uncomfortable a sunburn can be, regardless of whether we completely forget or fail to reapply our sunscreen.

Aloe vera is one of the most widely used at-home treatments for this kind of damage, and for good reason—just take a stroll down any grocery aisle. Aloe has several active ingredients that promote healing while lowering pain and inflammation. Aloe is also useful in hastening the healing process for burn victims, according to clinical research. Take one leaf and cut off the top and bottom layers of skin to begin enjoying the advantages of the aloe plant. Next, remove the leaf's two sides' spines and apply the cool gel inside to the damaged region.

One kind of succulent is the aloe plant. This indicates that it needs very thorough but sporadic waterings. It's time to water if the top ⅓ of the soil is dry! The ideal location for aloe growth is one with bright, indirect light. A combination of bark, perlite, rock, and potting soil is appropriate for the well-draining soil. It is necessary to plant in a container with drainage holes.

Honey.

One of the most traditional and adaptable medicines is honey. Honey has been used since the beginning of herbal therapy and has many beneficial effects, particularly for burn victims.

Honey is the complete package. Bees use floral nectar to produce a viscous liquid that has antibacterial and antifungal properties that also help fight inflammation. Honey attracts moisture and keeps the burnt area wet, which helps to avoid the formation of a scab. It's also well known that honey promotes quicker skin restoration.

There are a few key considerations when using honey as a self-medication for burn injuries. Burns ranging from first to light second degree can be effectively treated with honey. To ensure appropriate healing, you should contact or see your doctor for any burn that is more serious or moderate. Apply the honey to the afflicted region by first coating a sterile gauze pad. To maintain the dressing's hygiene, change it up to three times each day. Make sure you have medical-grade honey on hand in case you are hurt. The purity level of this product is higher than that of an ordinary store-bought bottle.

Lastly, and perhaps most critically, make sure the area is well cleansed before putting anything to a burn or injury. By doing this, you can make sure that there are no new issues that might impede your recovery or make your symptoms worse.

THREE TYPES OF HERBS TO HAVE ON HAND FOR EMERGENCIES

Cayenne

Circulation is enhanced by the stimulating hypertonic properties of cayenne. It can aid with sinus infections and open up airways.

When blood pressure falls, as it does during a heart attack or when a person is in shock, cayenne also aids in the dilation and relaxation of blood vessels. It is also effective in halting bleeding, both superficially and inside.

When animals are given cayenne anesthesia and don't wake up, they instantly awaken.

Yarrow

The plant's blooms and upper leaves, known as yarrow, are useful for halting internal or external bleeding.

Yarrow is a blood vessel constriction agent. It combines wonderfully with Shepherd's Purse to provide a remarkable clotting effect.

Lobelia

To clear airways, use lobelia. It works as a muscle relaxant and expectorant. It works wonders for allergies and asthma attacks, acting as an antihistamine.

Keeping gum weed and cramp bark on hand is also a smart idea, since they function similarly to lobelia.

A great way to reduce inflammation and help muscles relax is to combine lobelia with Brigham tea.

BOOK 10: PREPPER'S WATER SURVIVAL GUIDE

The human body needs water to exist for a few days, yet it can go weeks without meals. A natural disaster like a strong winter storm might cause a disruption in the water supply. It is also possible, albeit unlikely, that terrorist activities may contaminate a community's water supply.

By following a few easy steps, you and your family can get ready.

- In case of an emergency, keep some clean drinking water on hand.
- Know where to find water in your house from concealed sources.
- Recognize the possible sources of drinking water that are outside.
- Find out the proper way to filter water.

STOCKING WATER

Store enough water for each member of your household to last at least three days in case of an emergency. It is suggested to have a two-week supply of clean water on hand in case of a natural disaster or terrorist attack that contaminates your community's water supply. At least two quarts of water should be consumed daily by an averagely active individual, however this might vary. In addition to double the quantity of water kept during the summer, children, nursing mothers, and sick individuals may require more. Additional water will be required for hygiene and meal preparation. When getting ready for an emergency, it's advised to save at least one gallon of water per person per day. Included should be a sufficient amount of water for pets.

Glass, fiberglass, enamel-lined metal, and carefully cleaned plastic may all be used to hold tap water. Never utilize a container that has previously contained dangerous materials since tiny amounts can still be present. The finest containers are made of plastic, such one-gallon water jugs or soft drink bottles. Large plastic barrels or containers intended for food use that are specially designed to store water are also available.

Pour either tap water or well water treated with two drops of chlorine bleach per gallon into the clean containers. area the water containers in a dark, cold area after labeling and securely sealing them. Every six months, change the water. Keep the water supply on shelves rather than the ground when storing water in a basement. Stormwater that enters the basement during a flood might pollute the water that has been stored. Transfer the water that has been stored to a higher level if the floodwater level rises further.

In most retail stores, you can buy commercially bottled water. Look for an expiration date on the label. If none is provided, bottled water bearing the NSF or IBWA seal should have a minimum of a year on its shelf. Reviving the flavor of long-term-stored water can be achieved by repeatedly pouring it into clean containers to replenish its air content.

The freezer is another option for storing water for a long time. Frozen water helps maintain goods in the freezer frozen until power is restored in the event of an electrical outage. To prevent the container from shattering as the water expands when it freezes, leave two to three inches of empty space at the top. Since water expands by 9% when frozen, a one-liter bottle needs to have between 90 and 100 ml of headroom to prevent breaking. Regardless of the amount of air space offered, certain glass containers may shatter.

To keep the water safe and reduce exposure to microorganisms, it's crucial to use sanitary precautions after opening a storage container. Don't open more containers than are necessary at any given moment to lessen the possibility of contaminated water. Opened containers should be kept in a refrigerator at or below 40 degrees Fahrenheit, or 4.4 degrees Celsius, if electricity is available. Take special precautions to prevent bacterial contamination of the bottled water if refrigeration is not available and containers are kept at room temperature. Within a day or two, use water from unsealed containers.

Water rationing is seldom a good idea, especially for kids and the old. After consuming what you require today, attempt to discover more for tomorrow. By being less active and maintaining your body temperature, you may limit the amount of water your body needs.

INDOOR WATER SOURCES

Water is present throughout your house in a number of locations when your stored water supply runs out. The water in your hot water tank, pipelines, and ice cubes are examples of indoor water sources. As a last option, you can utilize the water in your toilet's reservoir tank (not the bowl), but only after purifying it using one of the techniques listed below.

Because they can store an average of fifty gallons of water, hot-water tanks are the finest source of indoor water. Before using the water in your hot water tank, disconnect the unit, switch off the power at the circuit breaker, or shut off the gas at the intake valve. To recover the water, place a clean bucket or pail beneath the tank's bottom drain. The drain could have a striking resemblance to an outside water faucet. Recall that the water originating from the tank may be quite hot. Open a hot-water faucet at one of your sinks and turn off the water intake valve, which is often found above the water heater. This offers a vent to allow water to escape the tank. Gas or electric should not be switched on to the hot water heater until the water at the main water supply valve and the water intake valve of the hot water tank are turned back on and the hot water tank is filled with water.

Turn on a faucet at the top of the home to allow air to enter the piping before using the water in your pipes. Next, get water from the lowest faucet in the home using a clean container.

Verify that the water in your house is safe to drink before using any indoor water sources. Close the main water shut-off valve in your home if you are experiencing a lack of water or sewage line breaks, or if you have heard that the water supply in your neighborhood is tainted in any other way. Look for the pipe rising through the floor at the lowest level of your home on each side of the water meter to find the main shut-off valve. There might be a shut-off valve on this line. The local water utility company will come find the main water shut off valve for you at no cost if you are unclear of its location. To avoid being caught off guard in an emergency, find the location of the water shut-off valve beforehand.

WHAT IF THE WATER COMING INTO THE HOUSE IS CONTAMINATED?

If there's any possibility that tainted water may have gotten into your house, never use the water in your hot water heater or pipes. While it's not advised to wash or take a bath in polluted water, it can be used to flush toilets.

Never try to use the water, even after purifying it, if floodwater enters the home from the outside or rises through the toilets or basement drain. Sewage, oils, industrial, and/or hazardous wastes might pollute the water.

You could have a pressure tank that can provide a small amount of water in an emergency if your water supply is from a well. You may still get water from the well if you have a generator and the electricity goes out. The water from the well should not be used for drinking, cooking, or bathing until the well is cleaned and disinfected if the wellhead has been tainted by floodwater.

Even if your house has a water treatment system, such reverse osmosis (RO), activated carbon, or a water softener, the water it produces might not always be safe to drink. The chemical makeup of the contaminant, its concentration, the kind and state of the treatment system, and the operational environment will all affect how safe the water is. For example, some solvents and other volatile organic compounds cannot be properly removed by a reverse osmosis unit.

Even while it may be beneficial to combine multiple different kinds of treatment systems, no one piece of equipment can completely eliminate all impurities. If activated carbon filtration and/or sediment filtration are not employed in conjunction, silt particles and/or chlorine may clog the RO membrane. Certain pesticides and organic solvents can also be eliminated when combined with RO in other treatment methods.

It can be in your best interest to stop the water from going through the treatment system, even if your water treatment system is equipped to clean polluted water entering your home. Replacing or cleaning parts of a polluted water treatment system might provide a significant risk of exposure. For instance, RO is efficient in eliminating Anthrax spores, but the unit's subsequent maintenance might be fatal.

Outdoor Water Sources

Some outdoor water sources can be used in an emergency if your interior water supply is depleted. Before being used, outdoor water has to be cleaned since it can include a range of microorganisms, including bacteria and parasites that can lead to illnesses including cholera, hepatitis, typhoid, and dysentery. Never utilize water that has the potential to be polluted by radioactive, biological, or chemical substances. The following methods of water filtration won't make the water safe to consume in this situation.

Potential sources of outside water include the following:

- Streams, rivers and other moving bodies of water
- Ponds and lakes
- Natural springs
- Rain and snow

When gathering water from an outside source, stay away from dark, pungent, or floating objects. Before consuming, water that has been acquired from outside sources must first be filtered and then boiled. Be advised that chlorine does not, or only partially, work to control Giardia and Cryptosporidium in surface water. It is normally preferable to boil these two bacteria away. A different approach makes use of filters designated as "absolute one micron filters," or filters certified to ANSI/NSF Standard 53 for "Cyst Removal" by an institution recognized by the American National Standards Institute (ANSI).

Rain or melted snow that drips from a home's gutters or downspouts can be collected in barrels or lidded plastic tanks. The water purification treatment specified below must be applied to the collected water before it is used.

WATER PURIFICATION TREATMENT

Once the water has been collected in one container, allow any suspended particles to sink to the bottom and then carefully transfer the clear water from the top into another container. To get rid of any last bits, filter this water with a fresh cloth or coffee filter.

Boil the water until it reaches a rolling boil and then let it boil for at least one minute to disinfect. If the water comes from a source suspected of containing Giardia or other protozoa, boil it for longer (5 minutes is recommended at 10,000 feet above sea level). It is not advisable to boil water for longer than is necessary since boiling will kill disease-causing bacteria present in the water but also concentrate non-volatile chemical contaminants.

Give the water at least half an hour to cool. Pouring the water back and forth between two clean containers will help to re-oxygenate the mixture. This will enhance the flavor.

Use regular home chlorine bleach to disinfect via chlorination. The bleach should only contain sodium hypochlorite at a concentration of 5.25% to 6% as the active ingredient. No more soap or scents should be added. Sodium hydroxide, which is safe to use in water treatment, has also been added as an active ingredient by a prominent bleach maker. For every gallon of water, add 16 drops (¼ teaspoon) of liquid chlorine bleach; for every 2-liter bottle of water, add 8 drops. Mix by stirring. If you don't have a dropper, you can measure the right amount of bleach using the following table.

- 8 drops = 1/8 teaspoon
- 16 drops = 1/4 teaspoon
- 32 drops = 1/2 teaspoon

Allow the water to stand for thirty minutes so that the germs can be killed by the chlorine. There is no assurance that this approach will work against all encysted protozoa.

You can use it if it has a chlorine-like scent. Add 16 extra drops of chlorine bleach per gallon of water (or 8 drops per 2-liter bottle of water) if it doesn't smell like chlorine. Stir, let it stand for 30 minutes, and then smell it once more. You can use it if it has a chlorine-like scent. Find another source of water and dispose of it if it does not smell like chlorine. Household liquid bleach is the only substance that should be used to clean water. It is not advised to use other chemicals, such as iodine or water treatment supplies available from surplus or camping stores, if they do not contain 5.25 percent sodium hypochlorite.

Pour the water several times from one clean container to another if, after disinfection, the taste of chlorine is too strong. This will reduce the amount of chlorine in the water and enhance its flavor by forcing some of the chlorine off as a gas.

BOOK 11: OFF-GRID LIVING ESSENTIALS

A unique combination of independence, self-sufficiency, and connection to nature can be found when living off the grid. But this way of living also necessitates a higher level of emergency preparedness. Off the grid residents need to have the knowledge and abilities to guarantee their safety and survival in a variety of situations because they do not have immediate access to municipal services.

OFF-GRID SOLAR ENERGY

Congratulations for deciding to embark on your road towards off-grid life! We've put together this comprehensive guide on off-grid solar system design and installation to help you with your project because installing an off-grid solar setup might be scary.

You'll find a thorough explanation of how to go off the grid using solar energy within, along with accurate calculations to help you size an off-grid system to meet your exact needs.

Benefits of Solar Power Systems for Off-Grid Living

The following are some advantages of solar power systems for living off the grid:

Total energy independence: You can produce, store, and utilize power at any moment with the aid of these technologies. Your home energy solutions become incredibly convenient and self-sufficient as a result.

Ecologically friendly: The sun is the best renewable energy source. Its energy may be used to create power, reducing carbon emissions. Solar solutions are also a sustainable option for off-grid living because they produce no emissions, noise, or pollution.

Cost-effective and long-term: The solar systems have very little ongoing costs and don't need to be maintained frequently. Also, the little annual utility payments justify the expensive upfront price.

Scalability: Solar panels and home battery systems can be increased at any moment. Its potential may be increased by adding basic panels and battery packs.

COMPONENTS

This is a brief summary of the components that your off-grid solar system will likely have. Selecting parts that are particularly qualified for off-grid use is crucial. For instance, the majority of grid-tie inverters are not set up to link to a battery bank.

Solar panels

Sunlight is absorbed by solar panels, which then transform it into DC (direct current) power. Although certain panels may be sold under the name "off-grid solar panels," this is a bit misleading. Panels used to be made to match the lower voltages of particular kinds of battery banks and charge controllers, however design standards have grown out of date due to advancements in technology. These days, marketing a panel as "off-grid" usually indicates that its wattage is lower than the industry norm, and most panels that carry this label are often of lesser quality.

We may now employ ordinary, mass-produced solar panels for off-grid applications thanks to MPPT charge controllers. Any conventional 60/120 or 72/144 cell solar panel would function perfectly, and installing full-sized panels would be the most economical choice if you have the room on your land.

Common solar panel sizes:

The dimensions of 60- and 120-cell solar panels are approximately 3.5 by 5.5 feet. The distinction is that half-cut cells, which are somewhat more effective and resilient to failure, are used in 120-cell panels.

Solar panels with 72 or 144 cells are around 3.5 by 6.5 feet, and the 144-cell panels also use half-cut cells.

While 72/144-cell panels are less expensive to install, 60/120-cell panels are easier to transport and provide more customizable design possibilities.

Common solar panel types:

Solar panels that are monocrystalline, or mono, are made from a single silicon piece. Compared to polycrystalline (poly) solar panels, which have cells composed of mixed silicon pieces, they are somewhat more efficient.

Because they can accommodate more solar in a smaller space due to their higher efficiency, mono solar panels are somewhat more expensive than poly solar panels. Both mono and poly solar panels will function equally well in terms of producing electricity, but a poly panel array will require more space on your home.

Batteries

the main component of solar systems outside the grid. Batteries are devices that store energy. Your battery bank may be used at any moment to power your gadgets.

Deep cycle batteries, which can be progressively depleted and recharged, are used in Off Grid solar systems. Solar batteries are usually designed to last you one night's worth of energy use, at which time they will recharge from the sun and complete one cycle of charging and discharging over the course of a day.

Several typical battery kinds seen in off-grid solar applications include:

Flooded lead acid batteries

Because the electrolyte is liquid and accessible by unscrewing the battery covers, flooded lead-acid (FLA) batteries are also known as "wet cell" batteries.

The electrolyte solution in flooded batteries evaporates while charging, thus distilled water must be added often to maintain the batteries full. Flooded batteries are only appropriate for people who have the time (and the willingness) to do routine maintenance checks on their battery bank once a month since they require regular maintenance.

We find that most people are unable or unable to adhere to the monthly maintenance plan that is necessary to properly care for FLA batteries, which are particularly prone to failure if improperly maintained. Unless you truly enjoy the concept of being hands-on with your system, their rigorous maintenance needs make them unsuitable for vacation houses, and we would not suggest them for full-time off-grid dwellings either.

Sealed lead acid batteries

Because the electrolyte compartment is covered to avoid leaks and toxic vapors from the battery, sealed lead acid (SLA) batteries receive their name.

Sealed batteries don't need to be housed in a vented battery enclosure and need less maintenance than flooded lead-acid (FLA) batteries. Because the contents of SLA batteries are sealed tight, they may also be installed in any direction.

Absorbent glass mat (AGM) and gel batteries are the two forms of sealed lead acid batteries.

In colder climates, AGM batteries outperform gel batteries and are less costly. Higher charge and discharge rates are also possible with them. In the majority of off-grid solar applications, they are the more affordable sealed battery alternative.

AGM batteries are less expensive than gel batteries, which are an older technology. They are less accessible than AGM and need more time to charge. Although AGM batteries are often more affordable, gel batteries do function better in high ambient temperatures, thus they can make sense in hot areas.

Lithium ion batteries

Since lithium Ion batteries have a longer lifespan than SLA cells and are often three times more expensive, the greater initial cost is offset during the system's lifetime.

Lithium batteries are the most practical choice if you're looking for a high performance battery that you won't need to replace for ten years. They weigh less, need no maintenance, and have quicker rates of discharge and recharge. Lithium batteries may also be started small and expanded as needed because they are modular.

Inverters

The inverter serves as the system's center hub, directing power flow among its many parts. An off-grid-specific inverter is required for solar power that is not connected to the grid.

Modern off-grid inverters are multifunctional and possess "smart" features. Many inverters come with integrated MPPT charge controllers. Some can start the generator in the event that the battery power drops too low, in addition to accepting generator power inputs. The "brain" for system monitoring is built into inverters, enabling remote system monitoring. Additionally, many inverters can interface directly with the built-in BMS (Battery Management System) of lithium batteries to maintain appropriate charge levels and provide battery bank information for your monitoring.

The low voltage DC electricity from the battery bank is transformed into 120/240V AC, the usual format used to power domestic appliances, by your off-grid inverter.

Modern off-grid inverters come with a number of clever features that let you control your system. A few instances are the ability to monitor and maintain appropriate charge levels via direct communication with lithium battery banks, remote monitoring, and autonomous generator starting.

Racking

the base upon which your solar array is built. The Ironridge XR metal rail system is what we advise. Racking is common to both off-grid and grid-tied systems. It's only a metal framework supporting the weight of the solar array; no extra equipment is needed.

Both ground mount and roof mount racking are functional solutions, each with advantages and disadvantages.

Charge controllers

The process of charging batteries is controlled by a solar charge controller. Charge controllers manage the voltage that solar panels generate, preventing your battery bank from being overcharged.

Off-grid systems typically use 48-volt batteries, however most solar panels provide more electricity than is needed to fully charge the batteries. Charge controllers shorten the time needed to fully charge the batteries by converting surplus voltage into amps. This maintains the charge voltage at an ideal level.

The anticipated lifespan of your battery bank is shortened by both overcharging and undercharging, therefore it's critical to select the appropriate controller and set the battery charging profile correctly.

Charge controllers may be classified into two primary categories: Maximum Power Point Tracker (MPPT) and Pulse Width Modulation (PWM).

We do not advise using PWM controllers, an outdated technology, in off-grid dwellings. They have fewer alternatives for appropriate solar panels and are less efficient. Remote telecom systems and other less demanding applications are better suited for PWM controllers.

The technology known as MPPT controllers optimizes the current flowing into the battery bank and is more dependable and efficient. MPPT controllers automatically modify the voltage to charge the battery bank as effectively as possible as the amount of sunshine varies throughout the day. We only incorporate MPPT charge controllers into our solar kits in order to satisfy the needs of permanent off-grid living.

Power center

The "brains" of the system, such as the inverter, charge controllers, monitoring system, overcurrent and surge protection, AC and DC inputs and outputs, and wiring to connect everything, are all contained in a pre-wired box called a power center.

Purchasing an off-grid inverter with built-in most of these functions, or a pre-wired power center, eliminates the laborious task of appropriately mounting and connecting several components together.

SOLAR SYSTEM DESIGN

Living Off the Grid implies you are alone in charge of producing your own electricity; in the event that your energy storage system fails to meet your demands, there is no backup source of grid power. Because of this, it's essential to consider every aspect that affects solar production while sizing the system.

Factors that impact off-grid system design

Prior to beginning the system sizing procedure, take into account the following:

Sun Hours

There are regions of the nation that receive more sunlight than others. You must ascertain the number of solar hours you receive in your location, which is a gauge of the length and strength of sunshine in your area.

Fortunately, the National Renewable Energy Laboratory's (NREL) solar insolation charts eliminate the need for guessing. Locate DNI (Direct Normal Irradiance) maps and record the average number of hours that the sun shines in your area. The majority of US locations receive four to five hours of sunlight each day.

The monthly maps show that there is a sharp decrease in solar hour availability throughout the winter. During the winter, your solar output will not meet your demands; your generator will have to fill the gap.

In theory, you could increase the size of your solar array to make it function throughout the dark winter months, but the cost would be astronomically high—think three times the cost of the system. Sizing your solar array to be active for the most of the year and letting the generator take over in the winter is far more economical.

Obstructions

You want to maintain solar panels clear of objects that may shade them since solar panels perform best in direct sunshine. Look for anything on your construction site that can obstruct sunlight reaching your panels, like as trees or chimneys.

Remember that throughout the winter, shadows lengthen due to the sun's descending arc across the sky. Ensure the absence of shadow at your construction site during the entire year.

Micro-inverters or power optimizers can be used to lessen the effects if partial shadow cannot be avoided. They won't, however, produce as much as an array constructed in direct sunlight.

Orientation

When solar panels are facing straight toward the sun, which follows the Equator's passage through space, they generate the greatest amount of power. Thus, you should face your panels directly south if you are in the Northern Hemisphere. Face them north while in the Southern Hemisphere.

Make sure you can face your panels in the correct direction when choosing a project location. To get the most out of solar panels, think about installing a ground mount away from obstructions if you don't have a suitable area on your rooftop.

System Voltage

There are several voltages available for solar batteries, including 6V, 12V, 24V, and 48V.

The reason a 48V DC battery bank is our recommendation is because it's the most economical and efficient choice out there. The increased amperage from the system at lower voltages will need you to purchase additional electronics and make additional cabling investments (the amperage doubles every time the voltage is decreased in half).

For an off-grid home, 48V is a more suitable choice. Typically, 6V batteries are used and wired in series to provide a total of 48 volts for the greatest performance.

DETERMINE YOUR ENERGY NEEDS

Three important variables need to be taken into account when sizing an off-grid system:

- Peak power demand
- Daily kWh usage
- Nightly kWh usage

Kilowatt-hour, or "KWh," is the standard unit of measurement for the amount of power used by your equipment.

Start by listing the power consumption of every equipment. Next, note the number of hours you want to spend using each appliance each day. The sizing procedure cannot proceed without this information.

Crucial! One kilowatt equals 1,000 watts. Before calculating your kWh, make careful to convert watts to kilowatts; otherwise, your results may be inaccurate!

What is your peak power demand?

What are the electrical loads that you will need to run? Will they all run at the same time, or can you rotate the loads?

Your peak power demand is your total wattage usage when you are running all the electrical loads you need simultaneously. By staggering usage of major appliances at different times, you can reduce your peak power demand and bring system costs down.

Figure out how many appliances you expect to run at the same time, and add up their wattage consumption. The total is your peak power demand. Make note of this number, as we'll be using it to figure out your inverter size.

What is your daily kwh usage?

Multiply the appliance's wattage by the total number of hours it will be used per day using the load evaluation worksheet that you completed. For instance, if a 1,500-watt dishwasher is used for 30 minutes a day:

A 750 watt-hour (1,500 watts times 0.5 hours)

In order to convert from watts to kilowatts, don't forget to divide by 1000.

Daily use is 0.75 kWh (750 Wh / 1000).

To get your daily kWh use, repeat this process for each appliance you plan to use and add them all up. Put that figure on paper in your memo.

What is your nightly kwh usage?

The electricity you use during the day is generated directly by your solar panels. Your appliances will function on saved energy when the sun sets and the solar panels stop producing electricity, thanks to the battery bank.

Add up all of the appliances you'll use at night and total them here, using the same procedure as before. Common gadgets that operate during the evening and overnight include your refrigerator, TV, and smartphone charger. It's important to take into consideration your inverter's self-consumption rating, which indicates how much power it requires to operate.

Off-grid houses that are well-designed can consume as little as 3–4 kWh each night; but, if you wish to operate power-hungry equipment in the evening, such as an air conditioning system, your consumption may be greater.

Add up the kWh you use each night and record the total in your notes.

Battery bank sizing

Now that we have the aforementioned numbers, we can finally start calculating the system size. First, let's talk about the battery bank. Its size needs to account for both continuous and peak consumption. We'll go over the calculations for an example off-grid setup with the following energy requirements for the sake of illustration:

Account for inefficiencies

The process of charging and draining the battery bank results in some energy loss. To compensate for the inherent inefficiencies in the charge cycle, we must leave a little bit of extra space in the battery bank.

Multiply by 1.05 for lithium batteries to take into consideration a 5% loss throughout the charging cycle. Since the sealed lead-acid batteries we utilize in our example are closer to 20% inefficient, we will multiply by 1.20 to make up for it:

1.2 × 20 kWh/day = 24 kWh/day

Days of Autonomy

To keep system costs low for off-grid residential systems, we advise sizing your system to finish one charge cycle each day.

Take the result from the previous step and increase it by the number of days you want the system to operate before you need to replenish your battery bank if you require extra days of autonomy.

For off-grid dwellings, this is not cost-effective, thus we will omit it from our example. On the other hand, remote industrial applications—like an array powering distant telecom equipment—would be better suited for it. If the system is not being monitored in certain situations, you might desire a few days or weeks of autonomy, which you can account for here.

24 kWh/day x 1 day of autonomy = 24 kWh

Depth of Discharge

Thus far, we have determined the amount of energy storage capacity required to provide the property with power for a whole day. But, if we construct a battery bank with that amount of capacity, we would be using up all of our batteries every day, which is bad for them because it will shorten their lifespan and void the warranty.

Depth of discharge (DoD), or the amount of battery capacity consumed before recharging, must be taken into consideration in order to effectively maintain the battery bank. 30% is suggested for sealed lead-acid batteries. The batteries will be severely overworked and have a shorter lifespan if you go above 50%.

Since lithium batteries can safely achieve 80% DoD, they are more forgiving.

For the sake of simplicity, we shall perform this calculation using a multiplier of 3. As a result, we use one-third, or 33.3%, of the battery bank capacity when we use our 4 kWh every night, which triples our energy storage capacity.

A 50% DoD is accounted for by a factor of 2. With the warning that deeper DoD results in shorter battery longevity, anywhere between 2-3 is acceptable.)

Note: Instead of using the daily kWh we used in the previous steps, utilize your nightly kWh amount here. In order to meet the nightly kWh requirement, we are estimating the amount of energy that must be stored in the battery bank.

4 nightly kWh x 3 (DoD compensation) = 12 kWh storage in battery bank

SELECTING BATTERIES

We have reached the desired 12 kWh battery bank capacity for our scenario. It's time to calculate the number of batteries we require.

Lithium battery

Kilowatt-hours, which we also used to calculate our nighttime energy use, are the unit of measurement for lithium batteries. Our battery bank math would be: 4 kWh x 1.25 (DoD compensation) = 5 kWh storage in lithium battery bank because our sample system is planned to use about 4 kWh each night and we can use lithium batteries to achieve 80% DoD.

Regarding minimum battery bank size and inverter compatibility, lithium batteries adhere to the same regulations. For instance, three batteries with a 5.4 kWh capacity each are needed for a system with a Sol-Ark inverter and Fortress battery bank, for a total storage capacity of 16.2 kWh. Again, this may seem excessive, but it's important to make sure the battery can withstand the inverter's charging amperage in a safe manner.

Sla battery

The amp hour is a new unit of measurement for battery capacity introduced by sealed battery manufacturers, so things can get a little complicated here. This continuously measures the battery current. The formula for amp hours is as follows:

- Watt-hours ÷ Battery Bank Volts = Amp-hours (Ah)

Since the previous phase showed us that we needed 12 kWh, or 12,000 watt-hours, of storage, we can divide that amount by the voltage of the battery bank to obtain our desired amp-hour capacity. Our goal for this system is to achieve a 48V battery bank, which is common for off-grid residential applications. Usually, this is achieved by wiring eight 6V batteries in series:

- 12,000 Wh ÷ 48V = 250 Ah minimum battery bank capacity

We haven't finished yet! Although we still need to confirm that it is compatible with the inverter of our choice, this is the bare minimum battery capacity required to meet our daily kWh use.

There will be a minimum battery bank size requirement for the inverter. Should you fail to achieve that minimal need, the inverter will use too much solar power and the batteries will receive too much amperage. This is an easy technique to quickly fry your battery bank.

All of our off-grid kits above 2kW, for example, require the battery bank to have a minimum capacity of 415Ah. As a result, the 250Ah value we computed earlier is insufficient to meet the inverter's requirements; a 415Ah battery bank is required.

Observe that this minimal criterion for inverter compatibility is exceeded with a nightly demand of 7 kWh/night at 30% DoD:

- 7 kWh x 3 = 21kWh battery bank
- 21,000 Wh / 48V = 437.5 Ah

As long as you stay below the 50% criterion, using a 415 Ah battery and increasing the depth of discharge in this situation is acceptable.

You would need to add a second string of batteries to increase the system's storage capacity to 830 Ah once you surpass the 415 Ah @ 50% DoD level.

Finally, keep in mind that a battery's amp-hour rating varies according on its rate of discharge. We need to know the amp-hour rating at a C20 discharge rate for off-grid solar applications.

C20 simply denotes a 20-hour battery discharge cycle, which is ideal for solar power. We finish one full charge cycle in a 24-hour period with a 20-hour discharge and a 4-hour recharge. Make sure that while choosing your batteries, you are considering the C20 capacity rating.

SOLAR PANEL ARRAY SIZING

Thankfully, determining the size of your solar panel array is much easier. Divide your daily kWh usage by the number of hours of sunlight in your location.

Note: Do not use the value from the battery bank calculations for daily kWh usage in these computations; instead, use the original figure. For us, that translates to 20 kWh every day.

20 kWh/day ÷ 4 sun hours = 5 kW solar array

From there, we must include a small amount of overhead to cover the panels' degradation rate and inefficiencies.

The performance warranty for solar panels states that their output decreases gradually with each passing year. Your solar panel's degradation rate is computed as 20%/25 years, or 0.8% production loss annually, if the performance warranty states that the panel would perform at 80% after 25 years. An end-of-life 400W rated panel would provide 320 watts at most.

Furthermore, solar panels are tested under perfect circumstances in a lab with regulated temperatures and no obstructions. Solar panels frequently produce a little less electricity than their wattage rating in the actual world because they are unable to meet these lab-tested parameters.

Due to these reasons, we budget additional solar capacity to ensure that, even after deducting system inefficiencies, we meet our output targets.

A decent headroom percentage to adjust for inefficiencies is 20%. For this, double the size of your solar array by 1.2:

5 kW x 1.2 = 6 kW solar array

Seems very straightforward, doesn't it? To meet our daily needs, we require 6kW, or 6,000 watts, of solar power. Based on the panel's wattage, you may then determine how many solar panels you'll need.

INVERTER SIZING

After determining the dimensions of the solar array and battery bank, we had sufficient knowledge to select an appropriate inverter.

This is the point at which your electricity demand peaks. The amount of continuous electricity that inverters can process determines their rating. Up to 8,000 watts of power can be used simultaneously by appliances if your inverter has an 8 kW capacity.

We calculated the peak power demand for our example system to be 6 kW, therefore we would search for an off-grid inverter that is at least that big. The inverter nameplate rating may be slightly higher than your peak demand; this allows for days of high usage.

WIND POWER

The energy contained in a moving mass of air is known as wind energy. The wind has been harnessed for human benefit, from the rudimentary wind-driven machinery used in ancient cultures to mill grain to the contemporary power generating equipment.

Wind is a cubic energy resource. The power available rises cubically with wind speed. This indicates that increasing wind speeds is crucial, and the taller the towers, the better. Going higher is the tried-and-true, dependable technique to boost a wind generator's output, regardless of the type of turbine or tower. Furthermore, mounting a turbine atop a short tower is the most typical error made while using wind power.

After the wind resource itself, a wind turbine's swept area is the second most significant component that affects energy generation. The collecting area is the circle that is "swept" by the blades. A small collector area cannot produce a significant amount of energy. According to Betz's theorem, the wind can only be harnessed for roughly 60% of its energy before its performance is actually reduced by excessive slowness. Well-built machines can do around half of that in the actual world.

Turbines can be categorized according to their generating mode, directionality, orientation, and other features. Wind turbines oriented horizontally (HAWTs) are the most popular and efficient type. Although they can seem appealing to the uninitiated, vertical-axis wind turbines (VAWTs) continue to fall short in terms of longevity and performance—both for the companies and the machines. Both downwind and upwind designs, in which the wind strikes the tower before the turbine, and the turbine before the wind reaches the tower, have the potential to be very successful.

One of three categories usually applies to generating gadgets. Permanent magnet generators (PMGs), which usually consist of revolving groups of magnets passing by fixed coils of copper wire, are used in the majority of home-scale turbines. Wind-field alternators, which generate electro-magnetism in the alternator's rotating portion by harnessing a tiny quantity of wind energy, are found in some older devices. By utilizing traditional induction motors and allowing the wind to propel them over their typical working speed, induction motor/generators convert from energy consumers to energy producers.

For wind-electric homes, there are three main types of towers. Although they can be built in extremely close quarters, freestanding towers are the priciest and may be the safest to install and maintain. Although tilt-up towers may be maintained and repaired entirely on the ground, their installation and operation necessitate a big open space. Fixed-guyed towers require climbing for installation and maintenance. They come in lattice and pole types that do not tilt. These are usually the least expensive and require a moderate amount of space to install.

There is much more to a wind-electric system than just the tower and wind generator. Transmission wire, electronic controllers, batteries for grid-interconnection or storage, metering, overcurrent protection, and other standard electrical parts are also necessary. Batteries may also be used for store or backup. Because wind energy production requires a whole system, all suitable components should be selected for compatibility and functionality.

WHY USE WIND POWER?

There are several reasons why people opt to use wind energy, but the main one is probably that they want to! Other more focused incentives include independence, dependability, finances, the environment, and society.

Our physical environment is greatly influenced by energy and the decisions we make regarding energy. Burning coal generates more than half of the electricity used in the US. Mountaintop removal, pit, shaft, and strip mining are the methods used to obtain this coal. The earth, water, and air are negatively impacted by each of these coal extraction techniques, in addition to the workers' and the locals' health. Both natural gas and nuclear energy, which have risks associated with their use and fuel extraction that could harm the environment, are additional significant sources of electricity in the country.

All energy-producing activities have an impact, but if properly planned and executed, wind energy and other renewables have a smaller total impact. Wind-electric systems do contain embodied energy, but the fuel is free and replenishes every day, and the long-term environmental effects are minimal.

Depending on the circumstances, wind energy systems may also be financially advantageous. Owners of Off Grid systems frequently discover that using wind power during the windy season is more economical than extending utility lines or running more generators. To compete with subsidized dirty energy, living off the grid will require a mix of a strong wind resource, high utility rates, and respectable incentives.

Independence was the cornerstone upon which our nation was built, and even now, particularly in rural areas, Americans still possess a strong sense of independence. People enjoy cooperating and supporting one another, but they prefer to be independent of governments, utilities, and charitable organizations. Instead, they want to take care of themselves. This goal can be achieved with a wind-electric system, giving you back control over energy production and expenses.

The main concern of several owners of wind-electric systems is dependability. These systems, particularly hybrid ones that combine batteries and solar panels, may produce electricity with a high degree of dependability that is frequently greater than that of the utilities. In terms of individuals, this can help persons with health issues, companies that depend on certain loads, and public service organizations go on during storms and power outages. From an alternative viewpoint, grid-tied wind-electric systems help maintain the stability of the utility grid by providing owners with electricity and allowing them to share any excess with their neighbors through the grid.

Finally, a lot of individuals want to use wind energy because they think it's cool. When you stop to think about it, a good number, if not all, of the purchases we make are motivated by our own interests, needs for fun, prestige, beauty, and interest rather than just money or "sense." All of that and more are provided by having a wind generator over your house, which also makes a powerful statement about your uniqueness, morals, and forward-thinking nature.

How to Use Wind Power

Three of the four fundamental types of wind-electric systems are frequently encountered. The most basic kind is probably wind-electric water pumping, which uses electronics to connect an electric pump and a wind generator. When remote water pumping is required in a location that is windy, this system type can be practical. However, there aren't many, if any, full system packages available for this use, and solar-electric and wind-mechanical (water-pumping windmill) systems are typically used to meet the need. Although direct heating systems are also an option, the energy they use is less beneficial because most end users want to utilize their systems for purposes other than just heating, and they also don't require heat all year round.

An off-grid wind power system that runs on batteries frequently has a backup generator and solar panel array. Due to the lack of availability or desire for utility electricity, these systems must supply all electrical energy. Because they must maintain a battery bank charged and because there will be periods of extra wind where the excess cannot be used, this implies that they will create a great deal more energy than is required.

Similar in setup to an off-grid system, a battery-based grid-tie system also features a grid connection and the capacity to "sell" excess energy to the nearby utility. Because of the battery bank, these systems have the benefit of offering a small amount of backup power during power outages. Because storms, when wind is copious, are generally the source of electricity disruptions, wind-electric backup systems typically perform admirably.

Because battery-free grid-tie wind-electric systems don't require batteries and typically run at a higher voltage, they are the least expensive, most effective, and most ecologically friendly systems available. Utilizing all of the energy produced, they create grid-compatible electricity to meet household needs and sell it back to the utility. The only negative aspect of them is that they don't have any backup protection because they can't function when there are power outages.

Depending on the size of the wind generator and the wind resource available, wind electricity can power most electrical loads. Systems for direct water pumping and heating are intended for use with particular loads. The majority of systems supply power for a range of uses, such as lights, workplace supplies, home appliances, and more. Small and domestic wind generators are also utilized by enterprises, ships, distant communication sites, navigational aids, and other establishments. In addition to powering your lights and appliances, you might be able to heat your house using a heat pump or direct electrical heat if your wind resource is plentiful. Since wind tends to steal heat from dwellings, and the windy season frequently falls during the heating season, this is typically a favorable combination.

Hydropower

When people think of pure, off the grid energy, solar power immediately comes to mind. There are, nonetheless, other choices. Not just the sun, but other elements may also be used to generate energy. We'll look at off-grid hydropower in this guide, which uses flowing water to produce energy.

A word of caution before we continue. It is your responsibility to find out from the local government whether using hydropower is permitted where you live.

What is a Micro Hydropower System?

You're in luck if your land has a river or even a tiny spring running through it. It is capable of producing power. Upon installing a microhydropower system, hundreds of kW of complimentary electrical energy can be obtained. This may be sufficient to run a sizable house or even a farm.

How does it operate? Basically, a waterwheel, pump, or turbine sits at the heart of a hydropower system. The water's movement moves this component, and the rotation converts to electricity. A tiny hydropower system also consists of an alternator or generator, which converts the rotational movement into energy. A regulator is in charge of the generator/alternator. The last component of this system, wiring, provides power wherever it is required.

It's vital to note that, although the waterwheel may occasionally be partially buried in the stream, the water is often brought onto the wheel, pump, or turbine via a canal or pipeline.

Turbines and generators are available together for purchase. If you purchase them separately, make sure to carefully match the horsepower and speed of the generator and turbine. The DC/AC problem is another one. Since the hydropower system generates DC electricity, your options are to either install an inverter to the system or purchase items that operate on this type of energy. The low-voltage DC electrical power will be converted into 240V of AC electricity via the inverter. In the case of hydropower, battery storage of the generated electricity is less feasible. Hydropower is typically much more cyclical than wind or solar energy. If you decide to utilize batteries, place them close to the turbine because low voltage electricity is difficult to transfer over long distances.

Since a waterwheel, pump, or turbine is an essential component of the system, let's briefly go over each and outline the benefits and drawbacks:

Waterwheel: A waterwheel is the original part of the system and has been used for a long time. Although they are still available, waterwheels are not advised. When it comes to producing electricity, their massive bulk and sluggish pace are less than ideal.

Pump: In contrast, no turbine produces as much mass as a pump. They are often inexpensive. They perform similarly to a turbine when their functions are switched around. Nevertheless, they require a steady stream of water to function satisfactorily and are more brittle and inefficient.

Reaction turbine: This incredibly effective kind of turbine generates energy through pressure as the water constantly pushes against its blades. Due to their high cost, reaction turbines are less prevalent in micro hydropower installations and more commonly employed in big hydropower systems.

The more economical propeller turbine is the lone exception. It has three to six blades and functions similarly to a boat's propeller. On the runner, the blades are positioned at various angles. A tiny hydropower system can make use of a Kaplan turbine, a very versatile variant of a propeller turbine.

Impulse turbine: Usually used in high-head systems, this kind of turbine has a very straightforward design. The runner, or turbine wheel, is moved by the water's velocity, which powers them. The impulse turbine has several subtypes.

Jack Rabbit turbine: This little turbine just needs one foot of water depth to operate. Make sure this is enough electricity for your home because the output is 100W.

Pelton wheel: The jet force concept generates the energy. The water enters a pipeline that has a nozzle-shaped, tiny exit aperture. The buckets on the wheel are struck by a stream of water that emerges from the nozzle. About 80% of the jet's rotational efficiency is used to revolve the Pelton wheel, and the buckets are shaped to maximize the impact's effect. A Pelton wheel is a great option for high-head, low-flow settings and comes in a range of diameters.

Turgo impulse wheel: This is an improved Pelton wheel. The jet is slanted and smaller, striking three buckets simultaneously instead of only two as in Pelton's instance. The wheel now moves twice as quickly as before. Turgo impulse wheels are renowned for their dependability and ease of upkeep. It is more compact and has fewer gears—or none at all.

How to Measure Hydrosystems Head?

What exactly do the terms "head" and "flow" mean? These are the things to look for when evaluating whether or not your property has a viable micro hydropower location. While a running water supply is undoubtedly a terrific place to start, it is not sufficient on its own. The water needs to descend, and you can learn a lot about its capacity to produce energy from its head and flow.

The head is the vertical distance that the water falls, to put it simply.

Water flow is the amount of water falling.

The efficiency of most micro hydropower plants nowadays is about 53%. Given the head and flow of your site, a straightforward formula may determine the output you will have for such a system.

System output (in Watts) = Net head (in meters) X Flow (in liters per minute) / 10

Let's talk about how these decisive criteria are measured. As you can probably guess, a higher head is better because it will take less water to fall in order to generate adequate energy, which means you'll just need to construct a smaller, less expensive hydropower system. A vertical drop of more than three meters is referred to as a high head. A low head is defined as anything less. Finally, a conventional micro hydropower system is entirely unfeasible with a drop of less than 0.5 meters. Under such circumstances, you will require a device similar to the previously stated Jack Rabbit turbine, which is fully submerged in water.

How can one approximate the head size roughly? Obtain assistance from someone and prepare a funnel, measuring tape, and a minimum of 2.5 meters of flexible garden hose or any other suitable tube.

From the penstock intake's most practicable elevation point, extend the tubing downstream. Assist your assistant in holding the hose's upstream end as near to the surface as possible while it is submerged and the funnel is inside.

Lift the downstream end concurrently until the water stops flowing from it. The vertical distance between the water's surface and your end of the tube should be measured. For that portion of the stream, this is the gross head.

Move to where you are now and ask your assistant to put the funnel there. Proceed downstream and carry out the process once more, taking measurements. This method of measuring should be followed until you arrive at the location where you want to place your turbine.

You may calculate the gross head available for your prospective hydropower system by adding up all of these measurements. If the figure is satisfactory to you, you could choose to hire a surveyor.

How to Measure Hydrosystems Flow?

What about flow, then? It is also measurable.

The simplest approach is the bucket method. Use some boards to dam the stream, then redirect it into a bucket or other similar container. Keep track of the time it takes the water to fill the bucket. The flow number is calculated by dividing the capacity of the bucket by the total time needed to fill it.

An even more sophisticated technique for measuring is done using a weighted float. Take caution not to use it in too deep or too rapid of a stream. Pack a weighted float (such as a plastic bottle half-filled with water), some graph paper, a tape measure, a yardstick, and a timer. Get someone to help you. The technique aids in estimating flow at the streambed's lowest point.

Choose a segment of the stream that has the most consistent breadth, depth, and straightest channel. Measure the width of the stream at its narrowest point.

At regular intervals, measure the water's depth while crossing the stream while holding the yardstick upright. A rope with specified increments can be stretched.

Plot the depths using the graph paper. This is the stream's cross-sectional profile.

Identify the areas of each part. To achieve this, get the areas of the triangles (area = base X height divided by 2) and rectangles (area = length X width). Repeat for each segment.

At least six meters upstream from the location where you took the width measurement, mark a point. In the center of the stream, release the weighted float. Calculate the time it takes the float to go downstream to the starting position using the timer. Use a smaller float if the one you have is dragging on the bottom.

You may now split the time in seconds by the distance between the spots. The flow velocity is given to you in meters per second. To obtain an even more accurate measurement, repeat this process many times. Determine the average speed.

Next, multiply this velocity by the previously measured cross-sectional area.

Lastly, multiply the outcome by the roughness factor. There are three distinct roughness factors for different stream channels: 0.6 for a streambed with many huge stones, 0.7 for a bed with tiny to medium-sized stones, and 0.8 for a sandy streambed. You will receive the flow rate in cubic meters per second as a final result.

Remember that the flow is frequently seasonal. To be safe, take the lowest average flow into consideration while designing your system. However, you may be limited in how much water you may divert depending on the season by municipal rules. In this instance, use the average flow rate from the time when electricity consumption was at its peak.

SELF-SUFFICIENT LIVING SKILLS AND PRACTICES

A degree or certification in off-grid living is not necessary to create a sustainable way of life. However, you want to be equipped with a few fundamental abilities to help you get through both difficult and routine tasks.

These are some basic off-grid abilities that all homesteaders need to possess.

Basic repairs.

When something breaks, someone needs to fix it. Eventually, you may save money by learning to identify issues early on.

If you utilize solar power to generate your electricity, be sure to clean and frequently examine your solar panel and batteries to detect any anomalies early on. Verify that the cables are securely fastened and undamaged, and verify sure your inverter is operating as intended.

To preserve the performance of your solar system, if you reside in a snowy area, you will need to brush any snow off your solar panel array using a soft brush.

Additionally, wind systems require routine cleaning and repair. Examine the concrete foundation beneath your wind turbine to identify any little fractures and seal them before they become more serious issues that might jeopardize the foundation's stability during the upcoming freeze-thaw cycle. Additionally, you want to inspect your little wind turbine's nuts and blades. Make sure there are no fractures or erosions on the leading edge of the turbine's blades, and tighten any bolts that have begun to come away.

Regular maintenance is also required for your pump and water source. Small problems, like changing a switch or mending a pipe, could be something you can do on your own. However, more serious problems with the submersible pump necessitate contacting your nearby well firm.

Depending on the type of septic system you possess and the suggestion from your septic installation, a septic system should be properly evaluated every one to three years. The interval between system pumping can be extended by routine care, such as keeping plants and trees away from the drain field and utilizing septic-friendly cleaning supplies.

It's important to keep leaves and other debris out of rain collecting systems, such as gutters and the rain barrel you use to store water. In addition, to prevent splitting or breaking of the barrel, pipes, or hoses when the weather turns cold, you should drain your rain barrel and disconnect the garden hose, even if you don't have a sprinkler system to blow out before the first deep freeze of the year.

You may learn how to perform small-scale home repairs, such as clearing up sink drains, touching up paint, and repairing a leaking faucet.

Build a fire

The aroma of a cozy, crackling bonfire is probably something you adore as well if you appreciate living off the grid. But fires are for more than just camping excursions when you live off the grid full time. You rely on them to cook your meals and heat your house, and a reliable fire may be really helpful.

Not only is building a fire a necessary bushcraft skill, but it's also useful at home. Gaining the ability to start a fire without the need for matches or a lighter will instantly give you a newfound sense of independence and security.

gardening.

Growing and gathering food is one of the primary motivations for starting a homestead. There's no feeling of self-sufficiency and preparedness like that after a food shop trip.

Plant hardier types that are heritage, non-GMO, and hybrid, and preserve the seeds to grow them again and again.

Gathering.

The ability to recognize the seasons and gather fruits at their peak is crucial, whether you're harvesting from your orchard or wild, native trees and bushes. You could believe that gathering food from the wild is something you can do whenever you want to if you're accustomed to picking fruit out of grocery store bins.

But nature doesn't operate that way. Fruit from plants that are grown in a natural setting in harmony with the seasons can mature over time; the range of dates might differ somewhat from year to year based on factors including rainfall, sunshine hours, and soil composition.

Spend time in the outdoors observing the growth of plants and their seasonal variations. Before long, you won't be watching the local news to learn the weather; instead, you'll be able to determine it from the way leaves change or from the behavior of local birds.

Food preservation

understanding how to cultivate and gather food also involves understanding how to preserve it. One of the most important homesteading skills while living off the grid is food preservation.

Creating a root cellar and knowing how to can and dehydrate food can help you to use your produce far into the short winter months. How comforting is it to know that your home food storage is your very own healthy version of the grocery store?

It is a reassuring sight to see shelves stacked with jars of fruits and veggies in every hue of the rainbow. You can feed your family all year long if you understand how to preserve food and take the proper canning measures.

Wildcrafting

The practice of collecting wild plants for food and medicinal is known as "wildcrafting," however other people only call food gathering "foraging." If there are any plants growing where you live, you may use some of them as additional food sources or to make herbal remedies.

repairing and stitching.

It's critical to acquire the fundamental sewing and repairing skills. The only equipment needed for simple tasks like hemming a pair of pants or sewing a button onto a shirt are scissors, a needle, and thread.

Cooking

Cooking is one of the fundamental homesteading skills and a component of food preservation. After all, your isolated off-grid residence is probably not going to be able to receive pizza delivery.

You can cook pasta and soup if you know how to boil water. Develop your bread-making abilities by learning how to create simple breads like tortillas and flatbread. This will help you keep your own sourdough starter going and produce fresh rolls every holiday season.

The practice of animal husbandry

The problem of sustainable living also includes animal husbandry. The majority of modern Americans do not know how to grow sheep for wool, hens for eggs, or a backyard cow for milk. However, as our culture shifts away from mass manufacturing and toward more sustainable living, these talents are being relearned and valued once again.

One more advantage of keeping animals? Garden soil that is rich in organic matter!

You may immediately use the dry pellets produced by goats and rabbits in your garden. Manure from cows, horses, and chickens is regarded as "hot" and needs to be aged before being applied to plants that have a high nutritional content in order to prevent burning.

First aid

The most important skill is first aid. In the event of an accident, what would you do if the nearest hospital is an hour away?

It might be a good idea to brush up on your emergency first aid knowledge if you're not a physician, nurse, or EMT. Purchase a how-to book, enroll in a Red Cross class, and assemble a couple of first aid kits. (Maybe one each for the vehicle, the home, and the barn?)

One last thing you ought to be aware of

There is always more to learn, regardless of your degree of expertise at the moment.

BOOK 12: PREPPER'S HOME DEFENSE

You've assembled a fully supplied prep pantry.

You've been saving clean water for your family with diligence.

You've provided alternate means of communication for your loved ones and yourself.

To put it another way, you're prepared for any SHTF (Shit Hits The Fan) situation!

or so you believe.

What happens if the amiable man next door is, in fact, an unfortunate zombie who neglected to prepare and now wants a share of your supplies?

Will the golden horde be able to defeat your home?

Here's how to fortify your home for SHTF (Shit Hits The Fan):

WHY IS IT NECESSARY TO FORTIFY YOUR HOME?

They claim that people's best qualities emerge after tragedies. But preppers are aware that they also attract the worst. When a disaster occurs, crimes like looting are all too common. And where do you suppose desperate people will go next, once all the grocery stores and restaurants have been destroyed?

Sherlock, no crap. Your stronghold.

Though that's not a very pleasant idea, it does illustrate that you should be preparing more than just the necessities in case you need to bug out or bug in. Moreover, you should work to strengthen your home's defenses against intruders.

THEN, HOW PRECISELY DO YOU DEFEND YOUR HOUSE?

Before TEOTWAWKI(The End of the World as We Know It), you should think about performing the following:

First Things First, Never Spill the Beans

The First Prepping Rule? Prepping shouldn't be discussed.

Nobody outside of your family should be aware that you have supplies, food, and water hidden away. You should prepare for everyone and his brother to come knocking at your door once news gets out that your house has everything someone could possibly need for SHTF.

That is, if they still possess the civility to knock.

You see, hunger has the power to transform even the most compassionate person into a vicious entity willing to harm you for the sake of a can of beans. You can choose to remain silent or suffer the consequences.

Camouflage Yourself and Your Home

In addition to keeping your mouth shut, you should ensure that your house and yourself blend in with the area. To keep people from suspecting you, you want everything about you to be as normal as possible. This is a simple method for

In order to safeguard your family as well as your survival supply store, you must:

- Don what your neighbors are wearing.
- Keep your purchases quiet.
- To keep curious eyes out of your windows, cover them with shades or drapes.
- Before throwing them out, rip the receipts.
- Refrain from sharing private content on social media.

Guard Your Perimeter

Keeping any potential assailants off your property should be your main objective when fortifying your house. But, you should keep them as far away as possible in case someone really tries to rob your home.

Making a defensive perimeter around your house is one of the greatest ways to accomplish this. Here are some pointers:

Tall Fences? Maybe Not

To ward off would-be thieves, some preppers advise erecting a 7-foot-tall wall or fence around your backyard. A large fence deters would-be thieves from breaking into your home, but it's also the clearest indication that you have valuables.

It is advisable to make your fence appear as innocent as possible if you plan to create one to protect your house and possessions. Your best option is a high timber fence or a strong chain link fence. Add barbed wire or spikes to the top of the fence when SHTF to prevent intruders from climbing it.

Landscape Defensively

Use your lawn's feature if you're fortunate enough to live in a home with one. Shrubs and trees offer more than just shade. You also get greater privacy with them.

However, be advised that people may utilize your plants as a means of self-defense. Thus, use landscaping as a clever way to strengthen your house. Ensure that your trees and hedges won't obstruct your view of possible invaders.

Don't just put regular plants, either. Choose the ones that are covered in thorns and prickles. Anyone brave enough to break into your home could suffer a great deal from these.

The following are some of the greatest plants you may grow for defense:

- Firethorn
- Century plant
- Bougainvillea
- Tomato porcupine
- Spanish bayonet

Don't Be Afraid of Open Space

You can have a clear line of sight if there is space between your wall or fence and your home's main structure. Because they can't hide their movements from you, trespassers become more exposed, which presents an opportunity for you to act.

Hang a Foreboding Sign

Another home defense tip is to put up notices around your property. These are really more of a deterrent. They might be completely stopped by posting signs that read "Quarantine area," "No Trespassing," or even "Trespassers will be shot on sight."

They'll ultimately choose a less challenging, threatening target.

Even if you don't actually own a dog, a "Beware of Dog" sign is still a classic. However, it might not be a good idea to announce that you have a dog during TEOTWAWKI. Your sad pet might end up on someone else's menu.

Set Up Cameras and Motion Lights

Believe that scanning your window for potential attackers will suffice when SHTF? Absolutely not. Security cameras and solar-powered motion-sensor lighting are worthwhile investments if you truly want to fortify your house. Distribute a few throughout your land. You won't be able to be ambushed by an invader.

Real-time footage from the security cameras will show their precise location and actions. They will reveal to you what your eyes are incapable of seeing. Having these cameras as additional eyes won't be a regret.

Regarding your motion lights, they will make it very difficult for burglars to remain undetected at night. When someone enters their area of vision, they are set off. to guarantee that they won't react to unintentional targets, such as random animals.

Fortify Your Home's Entry Points

What would happen if an opponent managed to breach your perimeter security?

What if they get through your door or window, climb your fence, turn away from your cameras, and ignore your signs?

This is when your home defense's next line of defense comes into play.

For the safety of your family and to keep attackers away, reinforced windows and doors are essential.

Improve Door Security

Zombies won't even try burrowing a tunnel into your basement or slipping down your chimney to grab your valuables after SHTF, especially the most ravenous and desperate of them all. The majority of them would just approach your door and force their way through.

Give them no opportunities. Boost door security by implementing these recommendations:

Put in top-notch deadbolt locks. Select one-inch deadbolts with a single cylinder rather than a double cylinder. When you're trapped in a fire, the latter can end up working against you.

If you use standard hollow-core doors—doors that even a 12-year-old can easily kick through—the best locks in the world will be ineffective. Solid wood or metal-insulated doors should be used in their place.

For your door hinges and strike plates, use longer set screws.

Put door jammers in place. Door jammers might buy you valuable time in the event that an entry attempt is made.

No deadbolt will stop an attacker determined to pry open your doors in the event that the world ends as we know it and chaos reigns. To strengthen your house, you should install additional reinforcements like flip locks, door barriers, and barricades.

Secure Windows

The best way to bring natural light into a room is via large windows. Unfortunately, if you don't safeguard your windows enough, other things will also be able to enter your home.

Probably a house's weakest component is its windows. They can be easily broken with a bat or even by using a large rock they find in the driveway, so no ingenuity is needed.

The best option if you want to fully secure your windows is to use bulletproof glass. Although it might break the bank, protecting your loved ones and your survival supplies from the horde of golden people is truly worth it.

If that's beyond your means, don't worry. You can also try these other window security techniques:

Use security film to reinforce windows. It prevents glass from breaking entirely, so to gain entry, an intruder would have to smash the window all the way through.

Place prickly shrubs or cacti under first-floor windows to hinder potential intruders' access.

Cover windows with burglar bars to keep intruders out even after the glass breaks. A quick-release mechanism is essential for enabling escape in emergency situations such as earthquakes and fires. This ought to positioned sufficiently enough from the window to prevent someone from reaching in to trigger the release and obtain admission.

Put locks on your windows. Most of the time, burglars take care to avoid drawing notice to themselves. When the locks force them to shatter the glass, they won't be able to contain their laughter for long.

Sandbags should be prepared to be placed behind windows in case armed robbers attempt to break into your house.

Set Some Booby Traps

Place booby traps outside and within your house to give your undesirable visitors an unforgettable surprise.

You can make a ton of things, from innocuous to lethal, to prepare your house for SHTF.

Though the deadly ones have a fierce appearance, they also perform their duties a little too effectively. We also don't endorse them because setting these traps is more likely to land you in hot water with the authorities than witnessing Armageddon.

Try the booby traps below as an alternative. Even though they are practically innocuous, they have the power to surprise the enemy. These traps will either force them to flee or occupy their time while you carry out your home defense strategy's next step:

- Tripwires
- Perimeter alarms
- Pit traps
- Corn flour explosive

Be Ready to Take Up Arms

A key component of any home safety strategy should involve self-defense.

These days, strengthening your home doesn't require you to be well-armed.

What you really need is a variety of reliable, non-lethal self-defense tools that you and your loved ones may use in any circumstance.

It makes sense to maintain a tiny supply of these things in each of your home's key rooms.

There are plenty of locations to hide your weapons, but it can be challenging if you have inquisitive children at home. For example, kitchen furniture, wall art, and fake outlets are excellent concealing spots.

However, common home objects and building materials can be used as makeshift weapons in an emergency:

- Fire extinguishers

- Golf clubs
- Baseball bats
- Crowbars
- Hammers
- Entrenching tool shovels

Designate a Safe Room

Zombies may still manage to get inside your house even if you have a well-thought-out plan in place to fortify it. Therefore, having a safe haven in case your home defense system fails is crucial.

Whatever name you give it—bunker, safe room, escape room, or panic room—your last resort needs to be strengthened as well to keep your family secure.

When creating your safe area, bear the following advice in mind:

You shouldn't wait to die in your safe room. It must give the same level of security as your external doors and contain a survival supply store.

It shouldn't stand out too much; it should fit in. You can make a walk-in area or a customized closet work perfectly.

It needs to have enough sleeping rooms for every member of your family.

Construct it at a convenient spot so every member of the family may escape swiftly.

If you don't have a lot of space, it can be difficult to outfit your home with a hidden living area. Therefore, you need be creative and figure out what works best for you.

Thoughts

The unprepared will be searching for food, water, and shelter when SHTF. To secure your survival and the survival of your family, you must be able to protect what is rightfully yours as a prep. And fortifying your house is one of the finest methods to achieve that!

Keep your prepping private, secure your perimeter, fortify your entry points, practice self-defense, and prepare a safe room in case things go wrong to strengthen your home defense while you still have time!

BOOK 13: RV CAMPING FOR SURVIVAL

In times of disaster and crisis, the ability to quickly respond and provide aid to affected communities is crucial. Using recreational vehicles (RVs) for disaster relief is an odd yet astonishingly successful approach that is gaining favor. Conventional relief efforts sometimes confront logistical obstacles. Explore how RVs have become into a useful tool for carrying supplies, giving shelter, and delivering relief during emergencies by reading on.

MOBILITY AND RAPID RESPONSE:

The portability and versatility of RVs makes them an excellent choice for disaster relief operations. RVs are meant to be self-sufficient and furnished with standard features including sleeping areas, kitchens, and restrooms. As a result, we can quickly dispatch our response team to the impacted locations without requiring elaborate logistical planning. RVs can travel across difficult terrain to get to isolated areas where help is needed right away.

EMERGENCY SHELTER AND ACCOMMODATION:

For displaced people and their families, providing temporary shelter is one of the most important demands when a disaster occurs. RVs can act as emergency shelters by providing a secure and comfortable haven for persons afflicted by natural disasters thanks to their self-contained living spaces. We can send out relief teams with our RV to disaster-stricken areas, making sure that people have access to basic needs like food, clean water, and shelter. Medical professionals can set up makeshift clinics and provide off-site care thanks to RVs' ability to accommodate them.

RESOURCE DELIVERY AND DISTRIBUTION:

Access to necessary supplies and resources may be severely restricted in disaster situations. With its storage capacity, the RV may be converted into a mobile distribution center that delivers food, water, blankets, hygiene kits, and medical supplies, among other necessities. We can access towns cut off from traditional supply chains by navigating through affected areas in the RV. Because of their versatility, RVs allow us to directly support individuals who are in need by providing resources.

COMMUNICATION AND CONNECTIVITY:

Coordinating rescue and relief actions during a disaster requires excellent communication. With communication tools like internet access, radio equipment, and backup phones, RVs can function as mobile command centers. Relief teams can coordinate operations, collect and share vital data, and provide real-time updates on the changing situation by using the RV as focal hubs for information transmission.

COMMUNITY SUPPORT AND REHABILITATION:

Apart from providing prompt comfort, we might also be essential in long-term healing and rehabilitation. The RV might be used as a mobile volunteer center, allowing people from the surrounding areas to drop off help in reestablishing communities. In order to allow volunteer teams working on reconstruction projects to remain close to the impacted areas and make a valuable contribution to the recovery process, the People Must Know RV can also be used as temporary accommodation.

In terms of delivering quick and efficient aid to impacted populations, the employment of RVs in disaster relief operations has shown to be revolutionary. During times of crisis, these vehicles are priceless due to their adaptability, mobility, and independence. RVs have shown to be a ray of hope and resiliency, doing everything from offering emergency shelter and distributing necessary supplies to fostering community solidarity and communication. In an unpredictable environment, adopting cutting-edge approaches like RVs for disaster assistance can have a big impact on reducing suffering and assisting in community reconstruction.

BEFORE DISASTER STRIKES

Fill Your Tanks

It's never a smart idea to park your vehicle with an empty gas tank, even though it's easier said than done. Sure, getting petrol on your way out of town is simple enough if you're not in a rush, but if severe weather is predicted for your area, finding a gas station will be the last item on your list of things to do when you should be prioritizing getting away from the storm. To swiftly get away from severe weather, you'll need a full tank. You should also have some extra fuel on hand so you can run your generator in case you get stuck somewhere without power. It's also important to make sure you have enough propane in your rig throughout the winter to keep it warm for a few days.

Make sure your tank holds enough fresh water to last you and your family for a few days. We typically fill the fresh water tank of our motorhome only halfway, even when we intend to stay somewhere with municipal water. The assurance that we could easily survive a few days off the grid is worth a little increase in gas expenditures.

Know your surroundings

When you arrive at a new place, spend some time getting acquainted with your surroundings. Find out whether there are any big trees or electrical lines close by that could be dangerous in the event of a tornado or earthquake.

It's crucial to ascertain the county you are in when you first park your RV because weather advisories are frequently distributed by county. For future reference, write down this information and save it in a convenient location.

If you ever need to escape your motorhome, know which routes will take you back to a highway and inquire around for the location of the nearest safe facility. We sought shelter in the RV park clubhouse during our tornado fright, but we subsequently discovered that there was a bathroom nearby with less windows. Even if it didn't matter to us in that particular situation, the seconds we lost by choosing the incorrect safe building may have been the difference between our family's safety and their safety.

Have a Reliable Alert Source

There are numerous weather applications and stations available, some of which are more dependable than others. Before the weather gets bad and you start worrying about contradicting information on the internet, do some research and identify a few reliable sites for weather alerts. My personal favorite is the AccuWeather App (available for Apple and Android), which is dependable and has real-time weather radar for tracking storms; however, NOAA Weather is equally reliable.

We recently bought a Midland NOAA Weather Alert Radio, which, provided we set it to our present location, will alert us to any extreme weather alerts in our area. The knowledge that we will be informed of all alerts even if we lose our internet connection offers us some little comfort.

We use the exact same radio, which is this one. You simply can't put a premium on peace of mind, even if it weren't reasonably priced, which it is.

STOCK UP ON ESSENTIALS

Making sure you have adequate food in your RV for a few days ahead of time is just as crucial as having enough water on hand. Having non-perishables and canned items on hand is especially important because any power outage could hasten the spoiling of food in your refrigerator.

Additionally, it's a good idea to carry a backup first aid kit in a tow vehicle if you chance to travel with one, as well as a first aid kit in your motorhome. Sometimes it's preferable to leave swiftly in your car and let insurance take care of any damage to your RV afterwards if you need to evacuate a place quickly. It will make you pleased to have a few necessities in your car if this ever happens to you.

Warm blankets, bottled water, fire extinguishers, and flashlights are some necessities that can be useful.

GENERAL RV CARE TIPS

Your RV needs maintenance that's similar to a combination of home and car care because it's both a vehicle and a dwelling. Every RV will have some common maintenance duties, much like every car has some maintenance requirements. Routine maintenance is necessary for all RVs, from oil changes to correct cleaning methods. Keep your RV in peak condition by adhering to these general maintenance suggestions.

Check the roof

While you're on vacation, your RV's roof shields you from the weather. To maintain it that way, you need to do some maintenance. Seeing the roof of your RV is not something you do frequently. Regretfully, the absence of problems does not imply their nonexistence. Your RV's roof may need to be replaced or repaired for several thousand dollars. Make sure your roof can keep the elements outside, where they belong, by using these guidelines.

Cover it up: Many RV owners find that it is just not feasible to use their camper full-time. For extra security when your RV is parked waiting for your next journey, cover the roof. You may protect your roof from UV rays and inclement weather by keeping it covered with an RV carport or a regular RV cover.

Clean often: Regular cleaning is the best preventive measure if you live in an RV full-time. The best method to prevent the dirt and grime that can collect in the cracks and crevices on your roof's surface and retain moisture is to have your roof cleaned on a regular basis for full-time users, or after vacations for those who use it occasionally.

Look for any leaks: Your RV's roof is significantly more intricate than your car's. Leaks can occur from vents, seams, seals, and air conditioners, among other things. Even before water gets into the car, a leak can do a great deal of harm. Before getting to the ceiling materials, water first soaks through the outside framework of your RV. Regularly check the air conditioning unit, skylights, vents, and edges of your roof to prevent costly damage that is hidden. Use a sealant that is compatible with the roof materials of your RV to fix leaks.

Store your RV properly

A lot of RVs are stored throughout the winter. Your RV's lifespan can be increased by properly storing it and selecting the right kind of storage. The three most popular kinds of RV storage are as follows:

The safest kind of storage is indoors, where you may keep your RV in a completely enclosed structure. Outdoor storage: Also referred to as open-lot storage, this choice exposes your RV to the weather. When keeping your RV outside, it's advisable to utilize some sort of cover.

Storage that is covered: This kind of storage offers the security of a roof and occasionally partial wall protection. Try to park your RV as far away from open walls as you can while using covered storage to prevent UV exposure.

Regardless of the kind of storage unit you select, there are a few maintenance chores that must be completed before leaving your RV parked for an extended amount of time.

Clean the awning

Awnings are a common addition to RVs, giving your pop-up camper or motorhome more shade. Cleaning and proper drying are essential before storing your awning, ideally after every usage. In the event that you are unable to give the awning a thorough cleaning, be sure to brush it off after each use to make sure all sticks, brush, and dirt are gone.

Inspect seals and slide-outs

If your slide-out is making a lot of noise, it's probably because you haven't done some necessary maintenance. Rust and corrosion can result from the accumulation of dirt and debris in your slide-out rails, which can impede effective sealing. Clean and lubricate the slide-out rails twice a year and whenever they produce odd noises or stop functioning correctly to prevent the need to replace worn-out rails.

Over time, window, roof, and door seals deteriorate naturally, which can let moisture inside your RV. Examine your RV seals for wear and tear every six months to keep them in good condition. Rubber seals should be lubricated to keep them supple and new. Examine each seal for degradation, water stains, and cracks. If seals are worn out or irreparably damaged, have them removed and replaced.

Check wheels, tires, and brakes

You stay secure on the road thanks to your tires and wheels. Before departing on any trip with your RV, make sure they are in good condition. Make sure your brakes, tires, and wheels can keep you and your family safe when driving by following these instructions.

Keep your tires shielded and clean: UV rays can quickly deteriorate tires, so you'll need to replace them before the tread wears out. Make sure you give your tires regular cleanings and apply a UV protectant. Use RV tire covers if your vehicle is being stored or left parked for a long time.

Check tire pressure and tighten wheel lug nuts: A catastrophic traffic collision could result from a tire rupture or lost wheel. Make sure your lug nuts haven't loosened before every travel. Verify the tire pressure and make any required adjustments. Road conditions can be risky when tires are either overinflated or underinflated.

Maintain the brakes on your RV: Brake maintenance should be done during spring maintenance jobs or before to your first trip of the season. Verify that the brakes are engaged and functioning as they should. Before you go, make sure you have the brakes checked and the wheel bearings lubricated. You can also consider getting new brakes.

Properly winterize your RV

Before being stored for the winter, all RVs, campers, and motorhomes in regions where temperatures may dip below freezing should go through a thorough winterization process. To keep your RV safe during the winter, adhere to these guidelines.

- Thoroughly clean the interior of your RV. This should involve taking out towels and linens from cabinets.
- Store and unplug your appliances.
- Empty and thaw the freezer and refrigerator.
- To keep rats out of your RV, look for any gaps or openings and seal them with metal dish pads that won't rust.
- To keep bugs out of your RV, stuff it with dryer sheets or cedar chips.
- Empty the water from the water heater, tanks, water lines, toilets, and other plumbing components in your RV, and then add antifreeze to the system.
- Keep the outside clean.

Examine seals for any damage.

- Shut all windows and doors and secure them.
- For outdoor storage, use tire covers and a fitting cover.

BOOK 14: CREATING COMMUNITY AND SHARING YOUR PREPPING

We are living in chaotic times, filled with everything from contrived disasters to never-ending economic instability, and governments trying to take away our liberties and right to live our own lives. However, there is a rising sense that we are about to enter some very difficult times. Even while we believe that everyone should be increasingly independent, there are situations in which having a strong support system or community is crucial.

We will examine why creating a support system of like-minded individuals is not only a need in an increasingly depressing world, but also a decision that goes beyond prepping. We'll discuss the advantages of creating prep communities, where to look for or start one, and what makes these groups work—trust, security, and cooperation.

The Benefits of Prepper Communities or Building a Survival Network

It should go without saying that having a survival network adds an extra layer of security to the advantages of having a reliable support system. It contributes to the assurance that you will not only be able to endure a crisis but also prosper when it has passed.

In case you're not persuaded, let us examine a few arguments for why you ought to think about creating your own survival network or preparedness society.

Sharing of Resources: No, we're not talking about living with a bunch of leeches or individuals who were brought up to rely on charity and government support. Those are the people you don't want around when things go south. But everyone in your community can share vital resources if you have a strong network of independent, resourceful, and like-minded people. In this manner, your network becomes its own supply chain in the event that supply chains collapse. Following a disaster, a well-organized community or survival network can establish trading routes or even pool food, water, and medical supplies to make sure that no one goes without food or access to healthcare.

Collective Defense: Let's face it, even in the best of circumstances, there seems to be an increasing amount of violence in the globe, as evidenced by the sharp rise in crime across the nation. When things go wrong, United, a prepper group, can assist defend you, ward off attacks, and keep off strangers who might want to pillage or steal your resources. Community defense tactics have occasionally proven to be extremely effective in maintaining safety, particularly in situations where looters and shady characters are trying to take advantage of chaos. During the LA Riots, for instance, rooftop Koreans banded together to protect their neighborhood's small businesses. They armed themselves, banded together, and saved themselves when law enforcement failed to protect them and their businesses!

Diversification of Skills: Although some of us may not like to acknowledge it, no man can possibly be an expert in everything. Having a broad group of people with a range of skill sets is one of the main advantages of creating a well-run prepper community or network, as this is necessary for survival in the event of an emergency.

Homeschool Co-ops: This is an additional fantastic benefit of creating a network of like-minded individuals if you have children. Being able to offer a personalized education is very tempting to families who live off the grid, and creating your own prep network with a homeschool co-op component can be a great way to escape the grid and government brainwashing.

SKILL DIVERSIFICATION: BUILDING A NETWORK TO SURVIVE!

We think this is one of the main motivations for building a network of individuals who can rely on one another in difficult times, therefore we want to go a little bit more into it. One of the best ways to make sure you survive and prosper in a crisis is to surround yourself with a varied group of individuals who are proficient in a variety of vocations and abilities. This is especially true in long-term survival scenarios where supply lines and access to basic necessities are cut off.

When thinking about the variety of abilities within your network, bear the following two points in mind:

Finding Community Competencies: Each person in your group should have specific knowledge and experience to contribute, whether it is in the area of medicine, marksmanship, trade skills, or wilderness survival. By recognizing and utilizing these abilities, you may create a community that is capable of overcoming obstacles that are insurmountable for any one person.

Educating One Another: Creating a network and developing your own skill sets at the same time is a terrific combination. Individual and group preparedness is improved by ongoing education within the community. You may also set up workshops and training sessions, allowing members to learn new talents or hone their current ones, depending on how comprehensive you want to go. When things go wrong, cross-training makes sure that everyone can contribute successfully.

We don't mean to seem condescending, but among the most neglected components of survival are mental preparedness and mentality. Allow us to discuss these briefly, as well as how your network may support you in addressing any mental health concerns that may arise during a crisis.

A Good Survival Network Can Provide Emotional Support

Undervalue the mental faculties in a survival crisis. Surviving a crisis will always require you to learn how to cope with stress, anxiety, and uncertainty; you need to be able to do this so that when things get hard, you won't break. By offering emotional support, prep networks can help make crisis preparedness or survival appear less daunting than going it alone.

Handling Anxiety and Stress: Everyone has seen the negative effects that stress can have on their mental health when things are tough.

Creating a Feeling of Safety: In times of crisis, knowing you're not alone helps you feel secure and confident. Knowing they have a community of people who share their ideals and are dedicated to mutual assistance typically brings comfort to community members.

Survivalist Qualities to Watch Out for: Survivalists frequently highlight some of the most important qualities when seeking for individuals who can benefit your community.

Finding or Forming a Prepper Community

Alright, so here's where the road meets the metal. Now is the ideal moment to begin building your network. A well-defined sense of direction and purpose is the cornerstone of a robust prep community. It takes more than just finding individuals to build a survivalist community—it takes finding the proper people. It is vital to have compatibility with regard to ideals, dedication, and abilities.

The initial stage of evaluating your needs is determining your survival priority. Your objectives will influence the mission of your community, whether it's protecting your loved ones, protecting your assets, or being ready for a particular threat.

So, how in the world do you go about making friends? Okay, let's examine a few choices. Let's start by examining current communities.

Joining Already-Established Communities.

Joining a random group of people should be done with some caution; if you don't already have a reliable group of people in mind, your best bet for creating a survivalist community may be to join an already-existing group. Here are a few locations to begin your search:

- Online Forums and Social Media Groups
- Local Meetups and Preparedness Events

Starting Your Own Community

Starting and creating your own community is, in our opinion, definitely the best course of action because it gives you the freedom to specify exactly what you want and select only those individuals who best fit your own aims and way of life.

The cornerstone of any prep community is effective communication:
Without a plan, things usually go south fast when the chips fall off. Because of this, it is essential to create a strategy that outlines how you would communicate and activate your survival network in the event of a crisis or disaster.

Creating Survival Network Communication Protocols In the event of a disaster

Effective communication can make the difference between life and death in times of crisis. It's time to set up your communication network and strategy so that in the event of a disaster, you can exchange vital information, coordinate swift action, and guarantee member safety.

A crucial aspect of preparedness is setting up clear communication protocols:

- Establish a communication plan
- Information Sharing
- Decision-Making
- Maintaining Morale

Good communication can save your survival network during a disaster. Clear protocols, frequent practice, and flexibility will help your organization manage crises more skillfully and improve survival rates. Making connections and working together with other preppers can be quite beneficial when confronting unforeseen circumstances.

BOOK 15: ENERGY INDEPENDENCE FOR SURVIVAL

Renewable energy is a vital but frequently disregarded aspect of disaster preparedness. Natural catastrophes have a catastrophic impact on communities, from hurricanes and earthquakes to wildfires and floods. In these kinds of circumstances, having a solid disaster preparedness plan can be quite helpful in guaranteeing people's safety and well-being.

Renewable energy has drawn a lot of attention lately as a clean, sustainable substitute for traditional energy sources. However, how is disaster preparedness related to renewable energy? Let's investigate this link in more detail and examine the advantages and benefits it provides off.

Off the grid operation of renewable energy sources, including wind turbines and solar panels, guarantees a steady supply of electricity in times of need.

THE ROLE OF RENEWABLE ENERGY IN DISASTER PREPAREDNESS

While conventional power grids may experience outages during natural disasters, renewable energy sources may offer a more reliable option. The relationship between renewable energy and successful disaster preparedness measures is highlighted by the following points:

RELIABLE AND RESILIENT POWER

Compared to conventional power sources, renewable energy sources including solar, wind, and hydro power are more resilient to natural calamities. Renewable energy systems can use power from decentralized sources, in contrast to conventional grids that depend on a centralized power generating and distribution system. them are less dependent on external power infrastructure that might be hacked during disasters thanks to its decentralized nature, which allows them to produce electricity locally. This guarantees a continuous supply of electricity, providing essential assistance to emergency services, healthcare facilities, and impacted people off-site..

SELF-SUFFICIENCY AND ENERGY INDEPENDENCE

A major benefit of using renewable energy in plans for disaster preparedness is the possibility of achieving energy independence and self-sufficiency. Communities can reduce their reliance on outside power sources during emergencies by producing their own electricity through the use of solar, wind, or hydro power. This self-sufficiency guarantees a steady power supply in both rural and urban areas and permits vital services to operate independently. Furthermore, by integrating renewable energy sources with energy storage technologies, excess energy can be delivered to nearby areas in need or saved for later use.

Off-Grid Power Generation

Because renewable energy devices can run off the grid, like solar panels and wind turbines, they are ideal for use in disaster situations. Because off-grid systems are not connected to the main power grid, they are less vulnerable to natural disaster-related outages. In times of emergency, when the main power grid may be destroyed or off line, off-grid renewable energy installations can continue to generate electricity, giving households and important services a dependable source of power. This capacity improves communities' overall resilience and makes it easier to respond to disasters in a successful manner.

The benefits of including renewable energy in plans for disaster preparedness.

Disaster preparedness systems that incorporate renewable energy off er various benefits. The following are some main advantages:

Enhanced Resilience: Power systems powered by renewable energy are more resilient to the effects of natural disasters, which speeds up the healing process.

Reduced Environmental Impact: Compared to fossil fuel-based energy sources, renewable energy sources emit fewer greenhouse gases and air pollutants, which lessens their negative effects on the environment both during and after disasters.

Cost Savings: By reducing reliance on pricey fossil fuels and possibly obtaining government incentives and tax credits, investing in renewable energy systems might eventually result in significant cost savings.

Community Empowerment: By integrating communities in the generation and delivery of energy, renewable energy projects promote a sense of shared responsibility and collaboration.

Communities can have a more comprehensive and long-lasting disaster preparedness plan by utilizing renewable energy technologies. This will ultimately increase resilience and guarantee better results during catastrophes.

To sum up, the utilization of renewable energy is essential for developing successful plans for disaster preparedness. Disaster recovery from natural disasters is facilitated by renewable energy sources, which also enable self-sufficiency and energy independence. In addition to being better for the environment, adopting renewable energy is essential for creating communities that are safer and more sustainable.

STRATEGIES FOR SUSTAINING ENERGY EFFICIENCY

The impact of natural disasters can be lessened and more robust energy systems can be built, despite the fact that they pose significant obstacles to energy efficiency. During and following natural disasters, the following are some crucial strategies to maintain energy efficiency:

Increasing the Resilience of Infrastructure.
The susceptibility of energy systems to natural disasters can be decreased by investing in robust infrastructure. Societies can lessen the impact and reduce energy losses by strengthening distribution networks, power plants, and transmission lines. Incorporating energy storage technologies can also improve grid stability and make up for intermittent renewable energy sources.

Embracing Microgrids and Decentralized Energy
Decentralized energy distribution and generation systems, or microgrids, are essential for maintaining energy efficiency both during and after natural disasters. In the event that the centralized system fails, these localized grids can function separately from the main grid and offer communities dependable power. Among the benefits of microgrids are:
Enhanced dependability: By dispersing energy generation across multiple sources and reducing the possibility of a single point of failure, microgrids can guarantee a dependable power supply.

Empowerment of the community: Microgrids promote resilience, self-sufficiency, and less reliance on flimsy centralized networks by enabling communities to produce their own energy.

Integration of renewables: By integrating renewable energy sources smoothly, microgrids can lower carbon emissions and improve sustainability.

Implementing Energy Conservation Measures

To lessen the burden on energy resources during natural disasters, energy conservation measures must be put into place. These actions may consist of:

Effective backup power usage: Promoting the use of energy-efficient backup power sources, such generators, can reduce the amount of energy used overall in emergency situations.

Knowledge and consciousness: Encouraging energy-efficient behaviors, like shutting off unused lights and appliances, can have a big impact on lowering energy consumption.

Energy audits: Long-term energy savings can result from doing energy audits in buildings and facilities to find and fix inefficiencies.

DIY ENERGY PROJECTS FOR OFF-GRID LIVING

PROJECT (BUILDING A SOLAR OFF-GRID POWER SYSTEM)

After designing the system and purchasing your solar kit, you'll need to learn how to install it.

You may either install the system yourself or contact a local installer.

People that live off the grid, in our experience, really enjoy taking the hands-on route and doing their own installations. To help those people prepare, we would like to provide a brief rundown of the installation procedure.

We are unable to provide detailed information on the wiring and system architecture because each off-grid system is unique and installation guidelines vary depending on the components in your system. To help you prepare for a DIY setup, we may offer a general plan that outlines each stage in detail.

Here is a summary of the steps needed in installing off-grid solar power to get you started:

BATTERY BANK

The battery bank need to be situated in a temperature-controlled, well-ventilated space, such as a shed or garage. Extreme heat reduces battery life, whereas low temperatures decrease battery capacity and slow down charging.

While batteries may function well in a very wide temperature range, 75°F is a reasonable goal temperature. It's advisable to construct an enclosure to protect them from the weather as the most crucial thing is to prevent extended exposure to harsh heat or cold.

It should be noted that lithium batteries are not as tolerant of lower (sub-freezing) temperatures and will have a temperature range that needs to be followed.

TIPS

- Aim for a room temperature of 75°F.
- Maintain a minimum of 1" of spacing between batteries to provide appropriate cooling.
- Check that there is adequate ventilation in the room because there can be some offgassing as a normal component of the procedure.
- Keep a notepad close at hand. To monitor the condition of your battery bank, you should test your batteries on a regular basis.

You may hook up your battery bank when the battery bank enclosure is ready. All that has to be done is follow the wiring schematics that are included in your owner's handbook.

Off - Grid Battery Banks are often made up of many smaller batteries connected in series. For instance, connecting eight 6V batteries in series results in a 48V total battery bank voltage. Every connection point will be shown in detail in the schematics, so just be sure you read the labels carefully and you should be set to go.

Some general guidelines for battery bank wiring:

The battery bank and the inverter should be kept as near as feasible. Voltage drop, or a tiny loss of efficiency as current passes through the wire, can occur with longer wiring lines.

Cables have to have uniform thickness and length. Unbalanced charging is caused by mismatched wires.

Every one to two feet, use zipties or clips to keep the wire neat.

To prevent damage, keep cables away from the ventilation system.

Verify again that all of the connections are properly torqued. Tight connections provide a fire risk and have the potential to melt breakers or terminals.

RACKING

Installing the racking that will keep your solar panels in place comes next.

Racking mounted on the roof

Your rafters offer a solid framework to sustain the weight of the solar array when used with roof mounts.

Finding your rafters and marking their arrangement is the first step. Searching beneath the edge of your roof for exposed rafter tails is the simplest method to accomplish this.

If your rafters aren't visible, you may find them by using a rubber mallet to tap on the roof's surface and listening for the sound it creates. If you hear a hollow sound, there isn't a rafter beneath. The sound will become a muffled, hard thump as soon as you identify a rafter.

A stud finder may be purchased, or you can detect the rafters in your attic and drill pilot holes all the way through to the roof. The drill holes will be hidden by the racking system flashings.

After locating your rafters, mark them with a chalk line (you can get one at your neighborhood hardware shop for around $5). These lines will indicate where your racking rails should be placed.

After marking the layout, flashings are installed by drilling them into the rafters at the designated positions. Each row of panels should have two rails supporting it, one at the top and one near the bottom, depending on how the flashings are placed. The specifications for the distance between your solar panels will be included in your setup instructions based on their size.

Bolting the racking rails to the flashings is not too difficult after the flashings are mounted. To ensure that the rails don't dangle over the edge of the array, you might need to trim them with a heavy-duty table saw.

Finally, grounding the array is required. Usually, the ground wire begins at the end of a rail, enters a junction box, and is then routed back to the power center through conduit.

Electrical safety rules require that all wiring be elevated off the roof's surface. To ensure that cables don't droop and come into contact with the roof, use zipties or wire clips to draw them snug.

Ground Mount Racking

To sustain the array's weight, a specific metal foundation must be built for ground mounting. Installing ground mounts requires additional money and time as you have to drill holes that are four to six feet deep and fill them with concrete to provide a solid platform for your system.

To do this, you will need to rent heavy-duty digging equipment. Since your cables from the array to the inverter must be buried, the digger will also be utilized to build a wiring trench.

The ground mount procedure resembles this:

Make fundamental holes.

After inserting the support beams into the holes, surround them with concrete.

Give the concrete time to harden.

Dig the wiring trench to the site of your inverter while you wait.

Install racking rails on the support beams once the concrete hardens.

Apart from constructing concrete footings and a framework to hold the array, the remaining steps of the ground mount installation procedure are similar to those of the roof mount installation described above: Attach the rails for the racking to the base structure, fasten the panels to the rail connectors, and run a grounding wire from the array's frame to the earth.

SOLAR PANEL ARRAY

In contrast, installing solar panels is rather simple once the racking has been constructed. The panels have mounting holes on the back that match up with the bearing locations on your racking. It just requires matching the panel mounting holes near the rail bearing point and bolting them down if your racking is arranged in accordance with your planset.

The connection point on the back of the panel will become unreachable once it is placed, therefore if you are placing panels on your roof, make sure to connect the wires before attaching the panel to the racking rail.

POWER CENTER

Your system's "brains" are located in your power center. It has AC/DC inputs, surge protectors, charge controllers, inverter, and monitoring system.

You won't need to bother about connecting the system's smaller parts because the power center is already pre-wired. Comparable to looking beneath your car's hood, you won't need to connect every part for an off-grid system to function, even though it's nice to know what's there. Installing the power center won't cause you any headaches because it has already been assembled.

Mounting the power center adjacent to the battery bank is recommended. As current flows through the wire, a longer wiring run between the inverter and the battery bank results in efficiency losses. As near to the battery bank as you can, mount the power center in a convenient spot. (The mounting location will be diagrammed out in your planset; this will be taken into consideration throughout the design phase.)

You may now wire everything together and connect it to the power center's inputs after it has been installed on the wall.

SYSTEM WIRING

It's now time to wire everything together after the racking, battery bank, solar panel array, and power center have been erected.

Giving general wiring guidance in this situation is difficult because each off-grid system will have different requirements due to the various component combinations required. Every wire and cable in your system will have a comprehensive wiring schematic included in your planset, complete with wire gauge, length, and connecting locations.

Pay special attention to the wiring parameters in your planset, as using the incorrect size wire may cause havoc with your battery charging calculations.

PROGRAMMING AND COMMISSIONING

Setting up your battery charging profile is the final step after all of your system's components have been installed and connected. To guarantee that your system maintains a healthy cycle of charging and discharging, this describes the voltage set points for the system.

You'll be able to enter these settings using a smartphone app or the power center's digital display; they will be described in your planset.

In programming, the following values are set:

Making ensuring you are utilizing the right size cable for your inverter/battery is crucial. If you don't, your inverter might not be able to handle heavy loads and might overheat, posing a fire risk. This should serve as a guide for selecting the appropriate cable size. If you have any more questions, don't hesitate to get in touch with our tech staff or a qualified electrician.

What is the size of your inverter?

What is your battery bank's DC voltage?

To find the maximum current for your cables, divide the wattage of the inverter by the voltage of your battery.

An Illustrated Calculation

Current (Amps) = Power (Watts) / Voltage (Volt)

Consider 1500 Watt inverter connected to the 24V battery bank.

(1500 W)/(24 Vdc)=62.5 A

Therefore, the maximum current that the cable must be able to handle in order to adequately provide the inverter with electricity is 62.5 A. 100A is the next larger size that is listed on the table.

To choose the right size cable for your application, refer to the above chart.

We can see from our example that 2/0 AWG cable would be suitable.

NOTE: Resistance via the wiring will cause a voltage loss over the wires for lengths greater than ten feet. It is advised that you increase the cable size to account for voltage loss if you must run wires longer than ten feet. Please feel free to call us if you have any questions concerning your application, and we would be happy to help you locate the appropriate cable.

PROJECT (BUILDING A MICRO-WIND TURBINE SYSTEM)

With the highest danger and failure rate, wind-electric systems are the most difficult renewable energy technology to build and manage. The majority of people would be better off employing or at the very least collaborating with a wind energy specialist to design, install, and maintain a dependable, long-term system, even though ardent do-it-yourselfers may see this as a challenge.

Assessing your site and yourself is the first stage in the process to find out if you have a good site, a good wind resource, and a favorable circumstance for a wind-electric system. Experience, knowledge, and "feel" are necessary for site assessments; most individuals overestimate their wind resource while underestimating the expense and difficulties of capturing it. Expert windsmiths use mapping, meteorological information, and firsthand knowledge of the area to decide how high to build the tower, where to put a wind generator on your land, and what kind of performance to anticipate.

The enjoyment and fulfillment of wind energy is a popular motivator. Making a compelling financial case for home-scale systems is often challenging, and the more compact the system, the more challenging it is. It may be possible to meet both your financial objectives and your subjective motivations when favorable conditions are met, such as high utility rates, a fantastic resource (average speed of more than 12 mph), and suitable incentives.

Make sure to consider the entire system when determining the cost of a wind-electric setup. The wind turbine is rarely the most expensive component of the package; it is merely one part. To create a functional system, you'll need tower and foundation, electronics, metering, transmission equipment, and more. Additionally, confirm that your long-term budget includes a sizeable line item for continuous maintenance. Your investment won't likely endure over time without this crucial care.

While the idea of having a dynamic wind turbine towering above your house and providing some or all of your electricity is alluring, it's advisable to approach the project with an open mind and a clear set of priorities. First and foremost, keep in mind that energy efficiency will end up being a more cost-effective purchase than a wind-electric system, for both the environment and your wallet. Start by reducing your energy use by investing in the most energy-efficient appliances and making prudent use of energy throughout your house.

Avoid the common pitfalls of wind energy:

Verify that your average wind speed, or wind resource, is sufficient for the job at hand.

Recognize the annual kWh output of each wind turbine you are thinking about.

Purchase a wind turbine that is big enough; a small collector won't yield much. The swept area serves as the collector.

Place your wind turbine high on a tower, far above any surrounding obstacles. You won't be pleased with the decisions you make on short towers and roof mounting.

To extend the life of your wind generator and ensure that you have electricity for many years, perform routine maintenance, ideally once a year.

BOOK 16: WILDERNESS SURVIVAL TECHNIQUES

Take into account these crucial wilderness survival skills to increase your chances of success:

BUILD A FIRE.

Start a fire with dry leaves, pine needles, or small pieces of wood to prepare food, keep warm, or scare off scavenging animals from your shelter location. To ignite tinder and kindling, use a firestarter or waterproof matches.

BUILD A SHELTER

You need a sturdy shelter for your comfort, safety, and survival when darkness descends or bad weather arrives. The following information will help you construct a shelter in the wilderness:

Identifying suitable shelter locations

It's important to pick the ideal spot for your shelter. Seek out regions that naturally offer protection, like the leeward side of cliffs or hills. Steer clear of low-lying locations that could flood. Make a note of any possible risks, such as loose rocks or dead trees. You can protect yourself from environmental risks and maximize the efficacy of your shelter by choosing a suitable site.

Building several kinds of shelters according to the resources at hand.

The materials that are accessible to you where you are will determine the kind of shelter you construct. A rapid and dependable shelter can be obtained by correctly setting up a tent or tarp. But in the event that you are without these things, you will have to make do with natural resources. Discover how to build lean-tos, snow caves, debris huts, and other shelters out of moss, branches, and other natural materials. Practice these methods ahead of time so that you are ready to construct a strong and secure shelter when the time comes.

INSULATING THE SHELTER TO KEEP THE WEATHER OUT AND PROVIDE WARMTH.

When constructing a shelter, take insulation into account to protect oneself from the weather and maintain body heat. To insulate the walls and roof of your shelter, add extra layers of materials such as leaves, grass, or pine needles. These strata will aid in retaining moisture and chilly air outside while trapping warm air within. Don't forget to allow adequate ventilation to avoid moisture inside the shelter.

You can find a safe and cozy haven in the woods with a well-constructed shelter. It will provide you with protection from the elements—rain, wind, and hot or cold—so you can sleep soundly and keep your energy up for the difficulties ahead. So, embrace your inner builder and create a shelter that serves as your haven in the splendor of nature.

BUILD AN OFF GRID CABIN

These instructions will help you construct an off-grid cottage that is both elegant and basic.

Selecting the Logs

Choose logs that are the length of your cabin plus four extra feet, with a minimum diameter of 10 inches. If you are cutting down the logs yourself, be sure to choose straight ones, debark them, and allow them to dry out for at least six months.

Laying the Foundations

Three typical types of foundations are pad, raft, and strip foundations.

The floor of the site is supported by raft foundations, which evenly distribute the weight of the entire cabin. Spreading one big concrete block at a time, it's usually less expensive and easier to lay than most others.

For cabins where there is no water logging and the site has a firm soil base, a strip foundation may be utilized.

The least expensive option is to use pad foundations, which require placing concrete piers at the cabin's four corners.

Raising the Log Walls

The two most popular are:

Butt and Pass

- Requires more pure strength and entails fastening the logs with big rebar rods.

Saddle and Notch

- Requires more accuracy, but will give you a more traditional-looking, well-constructed cabin.

Laying the First Logs

To ensure that the first two logs lie flat against the foundation, cut them in half lengthwise to give them a level base. Every one of these logs should sit atop the rebar that was placed into the concrete piers at the foundation phase.

Now to place the sleeping logs, completing the first layer of the hut. Using a scribe, make a U-shaped mark on the underside of the sleeper logs. Next, flip the log over and remove the notch using a chisel or chainsaw.

Turn the log over once the notches have been removed. It should sit snugly on the timbers that make up the sill below.

Fitting the Windows and Doors

After cutting the logs for the window or door with a chainsaw, place one of the logs for the lintels you set aside at the start of the project to guarantee structural stability.

Building the Gable Walls and Roof

Once the logs reach half the height of the gable wall, you will need to attach two purlin logs that run parallel to each other between the two gable walls.

As you construct the gable wall, make sure that each log progressively gets smaller than the previous one to form a triangle.

To connect the two peaks, use the last log you saved, the ridge log.

CREATE A HIERARCHY OF IMPORTANCE.

The "rule of threes" states that an average person can survive for three hours without shelter, three days without water, and three weeks without food. Anyone who finds themselves in a survival situation, whether as a hiker or otherwise, should carefully consider adhering to this guideline. Although these timings vary depending on the person and the environment, the rule of threes can serve as a framework to direct efforts in the field.

LOCATE A SOURCE OF PURE WATER.

In a survival situation, finding and gathering drinking water should be one of your top concerns because a human can only live for around three days without it. To boil the water, use a fire, iodine tablets, or a water filter. It's crucial to filter water in sufficient quantities to meet your demands for hydration, regardless of the method you pick.

SAFEGUARD FOOD AND WILD PLANTS

When you're in the outdoors, safeguarding your food supply becomes essential. What you should know about obtaining food and foraging in the woods is provided bellow.

IDENTIFY EDIBLE PLANTS, BERRIES AND INSECTS IN THE WILD

The profusion of food plants, berries, and insects that may keep you alive in the wild is one of nature's blessings. But it's important to know which species are safe to eat and which ones you should stay away from. Take the effort to become knowledgeable about the local plants and their tasty portions. Local authorities, internet sites, and guidebooks can all be excellent sources of knowledge. Always err on the side of caution when determining the edibility of a plant.

LEARN THE FUNDAMENTALS OF FISHING AND TRAPPING.

In survival conditions, adding protein-rich foods like meat and fish to your diet can be quite important. Gaining an understanding of fundamental fishing and trapping methods will improve your chances of obtaining food. Learn about the principles underlying several sorts of traps, including pit traps, snares, and deadfalls. Also, become acquainted with fishing techniques, such as the use of hooks, natural baits, and a basic fishing line. Before depending on these abilities in the wild, practice them in safe settings.

Recognizing the significance of wilderness food safety

While it's great to locate food in the wild, food safety should always come first. Here are some essential guidelines to keep in mind:

- Make sure to fully cook meat and fish to eliminate any possible bacteria or parasites.
- Before consuming wild foods, thoroughly clean and prepare them to get rid of any dirt, insects, or toxins.
- Avoid locations that are contaminated with pesticides or pollution when you go foraging.
- Keep an eye out for any dietary allergies or sensitivities you may have, and stay away from strange plants or insects that could make you sick.
- When taking resources from the environment, use moderation to protect the delicate ecosystem's equilibrium.

Though there is no shortage of food in nature, it is important to approach hunting and foraging responsibly and with awareness and ethical considerations. You can minimize your environmental impact while maintaining your own needs by acquiring food and engaging in responsible foraging.

Maintain impeccable hygiene.

Food and open wounds are two ways that bacteria, parasites, and diseases can enter the body. Maintaining hygiene and cleanliness will lessen the likelihood that you may become unwell and die. It's important to be healthy because even a few days in bed can swiftly exhaust your resources and lower your chances of survival.

Remain composed and weigh the circumstances.

Your mind is the most valuable survival gear that you possess. The human brain's amygdala will release stress hormones into the body to initiate a fight-or-flight response when you're in a life-or-death survival scenario. Resist the urge to act on your gut by pausing to gather your thoughts. This will enable you to reduce or eliminate needless risks and energy expenditure until you have a well-thought-out plan in place. Being composed is essential since mistakes that can be prevented can be fatal, particularly in remote areas.

Alert search and rescue personnel in the area.

Use visual and audible cues, like as smoke, whistles, and mirrors, to draw rescuers' attention and let them know where you need to go. If you're traveling alone, make sure your survival kit includes a solar phone charger to stay connected.

Examine your bushcraft abilities in advance of using them.

Even though you may already be able to make a bowline knot and navigate using the North Star, you should practice these abilities in a preparedness drill to see how well you do in real-world situations. While knowledge is important, practice can help you identify your advantages and disadvantages so that you can eventually get better.

Make use of everything available to you.

A survivor has to move fast and pack light, which means they have to carry multipurpose objects that are worth their weight in utility and gather food and water while they're at it. For example, a stranded traveler may not have a manufactured bug-out bag or first aid kit from which to gather supplies or tools needed for a comfortable survival. Situations such as these call for creativity and strong problem-solving abilities. A plastic bag, some paracord, and some duct tape, for instance, may be the main components of a shelter that protects you from the rain and wind.

Recall that this is only the start of your survival quest. Regular practice and improvement of these abilities is vital. Maintain your curiosity, keep learning, and look for chances to broaden your understanding of outdoor survival.

Above all, when it comes to your outdoor experiences, always put preparedness and resourcefulness first. Although the outdoors can be breathtaking, it requires caution and respect. As you make your way through the stunning but unpredictable world of outdoor exploration, your preparedness, adaptability, and optimistic outlook will be your best friends.

Now venture forth with courage, embrace the wild, and let your spirit of adventure to lead you on life-changing adventures.

BOOK 17: COMMUNICATION IN CRISIS

Following a significant disaster, getting in touch with loved ones is crucial in order to find out how they're doing and to make sure they're okay. However, for a number of reasons, communication can be particularly challenging during disasters:

1) The physical infrastructure supporting communications, such as switches, fiber optics, and phone lines, has been damaged.

2) The lack of power that powers communication devices.

3) There are so many concurrent users attempting to communicate that the communications infrastructure gets overloaded and unable to handle more users.

Your objective is to create a communication strategy that reduces bandwidth use and makes advantage of alternate channels in each of these scenarios.

COMMUNICATION OPTIONS IN A DISASTER

Landline Phones

An old-fashioned landline phone that is based on copper wire is frequently quite reliable in many types of calamities including floods and power outages, even though it can be challenging to find an actual landline phone in the age of the internet and voice over internet protocol (VOIP) digital communication. Because they run off their own low voltage power system, landlines continue to function even during power outages. Your best option while utilizing a landline is to have a wired phone (a non-cordless phone that runs on batteries or a wall outlet).

Making a call to your out-of-state acquaintance can assist in avoiding crowded and overburdened local phone lines. Additionally, even if you do not immediately hear a dial tone, when you pick up your landline phone, wait a few minutes with the receiver off the hook. If the lines are still intact, it may take a little while for the system to connect and give you a dial tone. Available lines are assigned in the order that requests are received.

Text Messages: Because there are fewer wire lines to break in cellular phone systems, they have shown to be more dependable than many landline phone systems during certain calamities like earthquakes. Nevertheless, during emergencies, cell towers frequently experience oversaturation of users. Because text messages use a lot less bandwidth on the already overburdened cell phone infrastructure, they have a higher chance of being delivered than voice calls.

After a disaster, cell phone batteries will soon run out of power when used extensively, so be sure to always have a portable battery backup (power bank) and power cord on hand. In a disaster, a car charger/adapter can be quite helpful because it can power your phone for days using the battery in your car.

Keep in mind that cell towers require electricity, backup batteries, and/or generators in order to operate. In the hours following a significant earthquake in 2011, residents in Christchurch, New Zealand, were overjoyed to have cell phone service; however, as generators ran out of fuel, the cell towers went offline.

Email: If you have internet access, email can be an excellent way to communicate during a disaster. Popular email services like Yahoo or Gmail are cloud-based, which lessens the possibility of a local email server failure that you could face with a work-based email like Outlook. Text-based emails (no photos) also take very little bandwidth.

Two-way radios

Two-way radios are the mainstay of public safety agencies' disaster communication strategies because they offer incredibly dependable communication when all other avenues have failed. The majority of two-way radios operate by speaking with nearby radios directly. This indicates that, under perfect circumstances, their range is only a few miles, or even less if you're in a metropolis with a lot of structures. You may be able to communicate with any other radio as far as your eye can see for many miles if you can reach the top of a hill and your range extends to line-of-sight. Most big box stores and internet retailers sell inexpensive two-way radios that are frequently used for recreational purposes, like the General Mobile Radio Service (GMRS) and Family Radio Service (FRS) radios. Although these radios can be purchased without a license, in order to use the increased transmission strength on GMRS frequencies legally, an application for an FCC license and the corresponding cost must be submitted. If you decide to utilize an FRS/GMRS radio, make sure you familiarize yourself with its settings and limits by using it prior to a disaster.

"Ham" Radio

In addition to being able to communicate between radios, ham radios can also have a wider transmission range thanks to the use of "repeaters," or intermediary relay radio stations. These repeaters are frequently positioned in advantageous areas, including the tops of mountains or large buildings. This makes it possible for a compact, handheld ham radio to successfully communicate over a whole local area. Other ham radios operate on lower frequencies, which allow their messages to reverberate around the nation or the world by reflecting off the atmosphere. A license is necessary to operate a ham radio, and it can be obtained by passing an FCC test that tests your knowledge of radio communication laws, physics, and concepts. Even with this prerequisite, ham radios remain one of the most dependable means of communication in the wake of a disaster!

BOOK 18: PREPPER'S FIRST AID MANUAL

Emergencies and disasters are regrettable but unavoidable aspects of life. Anybody can encounter an unforeseen but dangerous circumstance that they are unable to avoid, regardless of their age, place of residence, or line of work. People can make plans for natural catastrophes and medical emergencies, even though they can't always be prevented. People can reduce the likelihood of harm and death during and after dangerous incidents by being prepared. People can assist others in a crisis by preparing for medical emergencies and natural calamities.

First Aid Kits

An essential tool to have when getting ready for any kind of emergency is a first aid pack. It is a medical kit filled with items like thermometers, antiseptic wipes, antibiotic ointments, and anti-itch lotions. They also contain supplies including gloves, gauze, tweezers, and bandages. Depending on their intended usage, these kits come in a variety of sizes. Individuals may keep a fully equipped first aid kit in their automobile to address any injuries they may sustain while traveling. Additionally, a person should always keep a kit in their house, out of children's reach but still accessible to all adults. The family's emergency supplies can also be kept in storage along with a backup kit. These need to be routinely inspected in order to replenish any utilized items and get rid of any expired ones. This ought to be carried out ideally every three months.

Natural Disaster.

Examples of sudden and dangerous natural disasters are earthquakes, storms, floods, and fires. Making an emergency kit with essentials like nonperishable food, water, a torch with extra batteries, and blankets is a good way to start getting ready for one of these disasters. One must prepare for and respond to natural catastrophes differently because they are all unique and damaging in their own ways. Families ought to review the most typical natural disaster categories in their region. Although house fires can happen anywhere, certain regions may also be more vulnerable to earthquakes, while others might see tornadoes. If you're outside during an earthquake, it's preferable to stay in open areas. If someone is inside, they should hide under a strong desk or table that is not near any windows or any things that could fall on it. They ought to shield their heads as well when working beneath the desk. During a hurricane or other high-wind natural disaster, people should stay indoors, stay low, stay close to the center of the building, and stay away from windows. The same goes for earthquakes: they should take precautions to avoid hurting their head or neck. Having a prearranged escape strategy is important since staying home during fires and some other disasters can be fatal. At least two viable escape routes from each room should be included in the plan, which should be created with the assistance of other residents of the house. It is important to practice the plan so that everyone is aware of what to do. A meeting spot where every household member can congregate following a successful escape from the house is another item that needs to be planned ahead of time. A person who resides in a different city should also be assigned as a family contact following a significant event.

Medical Emergencies.

Medical crises, such as heart attacks, strokes, or severe injuries, are grave and frequently life-threatening situations that need for quick attention. It is not necessary to be a nursing practitioner or student to identify the early indicators of a possible medical emergency. These symptoms could include vomiting blood, breathing difficulties, or chest pain or pressure that lasts longer than two minutes. Enrolling in a CPR and first aid course can help someone get ready for medical crises. A list of emergency contacts should be kept visible and immediately accessible, or it should be programmed into a phone. These numbers ought to include the poison control center, the patient's physician, and the pediatrician in your household. It's a good idea to program a contact person's name and phone number into your phone, preferably someone who knows your medical history. This is to assist first responders in the event that the injured person is not responding; it should be labeled as "In Case of Emergency," or ICE.

Medical Supplies and Equipment for First Aid Kits

Emergency conditions and procedures are the main focus of first aid instructions, particularly those provided by paramedics. It's all about knowing what to do if you see blood oozing, missing parts, or stopped breathing. All of that knowledge is helpful, but the most useful first aid advice relates to common injuries, such as those sustained at a child's birthday party or at a company picnic.

The recommendations to learn CPR and dial 911 should not be disregarded. However, it's helpful to know how to treat minor injuries using first aid.

PUT AN END TO A BLOODY NOSE.

Bloody noses sometimes occur unexpectedly (and you should definitely report them to your healthcare professional), but most bloody noses are treatable, typically with digital trauma. That's just picking your nose. Inform the medical professional if you did not do anything to traumatize your nose before it begins to bleed. Otherwise, avoid sticking your fingers up your nose.

TREAT A CUT FINGER

The way a cut finger is treated is not unique. This first aid tip works just as well on a chopped nose, split earlobe, or ripped toe as it does on a pinky finger. If blood is seen flowing on your home's floor, it's probably coming from your fingers. Of course, there are also thumbs.

HANDLE A SPRAIN

Everyone eventually suffers a twisted ankle, even if they're not crawling on rocks or sliding into second base. Taking out the trash or playing Wii can cause a sprained wrist. You should be aware of how to care for a sprain as a responsible adult.

TAKE OUT A SPLINTER

You have to know how to take out a splinter from as early as kindergarten. Splinters are a common part of growing up, from playground equipment to trees and garbage. To assist prevent an infection, study how to do it correctly in case you missed any of the small details.

PUT AN END TO DIARRHEA

Even the sharpest first aid instructor made the mistake of failing to include first aid advice on how to halt diarrhea in the lesson. Should you have trip plans outside of your zip code, you may want to prepare yourself for the inevitable stomach turns you'll have. You'll probably need these tips at home too, since not all upset stomachs are caused by terrible bugs.

Address Nausea

It makes sense that if something is exiting one end, it must be exiting the other. If determining the reason of nausea is the first step, there isn't much more you can do to prevent vomiting. Still, every little bit counts. You ought to be knowledgeable about treating nausea.

Head Lice Killing

After taking a bath, you shampoo your hair. You mean there's no possible way that you could have head lice? False. A clean head of hair is the ideal habitat for head lice. That it's not the end of the world is the good news. Although they are really disgusting, head lice aren't very dangerous.2. You must be able to eradicate head lice.

Handle insect bites

Not all animals that bite are head lice. There are countless little biting insects in the world. There are a lot of them in your home at the moment. The most frequently utilized first aid tip of all is how to treat bug bites, aside from cuts on fingers.

Handle a Burn

It can seriously destroy the cookies to touch the rack in a hot oven. Fortunately, you are more than capable of managing the injury on your own. Put your finger under cold water after turning off the computer. Return in ten minutes to see how a burn is being treated.

ABCs of First Aid

You should be familiar with the ABCs of first aid if someone is unconscious or unresponsive: airway, breathing, and circulation.

Airway: Opening someone's airway should be your first step if they are not breathing.

Breathing: If someone's airway has been cleaned but their breathing is still not occurring, administer rescue breathing.

Circulation: To maintain the person's blood flowing while performing rescue breathing, give them chest compressions. Verify the person's pulse if they are not responding. Give them chest compressions if their heart has stopped beating.

A simpler version of the ABCs is:

Awake? Make an effort to wake the person if they are not awake. Make sure someone is dialing 911 if they don't wake up, then go to the following action.

Breathing? Rescue breathing and chest compressions should be administered if the person is unconscious and not breathing. Proceed to the subsequent action.

Continue care: When you dial 911 for assistance, heed their advice or carry on with your care until an ambulance arrives.

Some first aid courses also include D and E:

D might stand for automated external defibrillator (AED), lethal bleeding, or disability assessment. An AED is a medical device that shocks the heart to induce a heartbeat.

E can stand for: Examination, which is the process of looking for indications of a wound, bleeding, allergy, or other condition once a person's respiration and heartbeat have been confirmed.

First Aid Kit List

Many pharmacies and department stores sell first aid kits, but you can also construct your own. One should be kept at home and in your car.

A standard first-aid kit need to include:

- Adhesive bandages in multiple sizes and shapes

- Gauze pads in multiple sizes
- Compress dressings
- Adhesive cloth tape
- A roll of gauze
- Gloves
- Antiseptic wipes
- Antibiotic ointment
- Hydrocortisone ointment
- A breathing barrier for performing CPR
- An instant cold compress
- Baby aspirin
- Tweezers
- An oral thermometer
- An emergency blanket

The best approach to know what to do in the event that you, a loved one, or even a complete stranger has a medical emergency is to obtain proper first aid training.

Knowing the ABCs (airway, breathing, and circulation) and how to perform CPR is helpful even if you haven't received official training.

It's usually preferable to try first aid care rather than do nothing. Taking prompt action can occasionally save a person's life.

Book 19: Urban Survival Strategies

Being ready for emergencies is more than just a precaution in the uncertain world of contemporary urban living. Imagine a situation where the busy metropolis you call home devolves into anarchy out of nowhere, and the conveniences and luxuries you take for granted on a daily basis disappear in an instant.

We'll go into urban survival strategies in this guide that seasoned military veterans and expert preppers suggest to get through the perilous waters of a metropolis in crisis. So buckle up, because these tactics could mean the difference between thriving and just surviving when times are hard.

1. Develop Adaptability and Resourcefulness

Your ability to adjust to quickly changing situations is critical for urban survival. Enrolling in survival courses can help you learn how to do this quickly by fostering your creativity and decision-making abilities. These courses teach you how to navigate intricate urban surroundings and make the most out of scarce resources.

As water molds itself to the pitcher, so the wise adjust to their surroundings.

2. Prioritize Communication

In an emergency affecting the entire city, communication must be trustworthy. Invest in a communication tool that will continue to work even if regular networks go down. Satellite phones and two-way radios can help you stay in touch with emergency services and your family by bridging the communication gap.

The truth is that knowledge is the essential to survival, and its greatest advantage is that it is weightless. Ray Mears

3. Master the Art of Self-Defense

Beyond just being a useful tool, self-defense abilities may be essential to survival. Enroll in self-defense classes to give yourself the skills and information you need to defend your loved ones and yourself.

Furthermore, think about lawfully carrying self-defense items like pepper spray or personal alarms, which can save vital seconds in an emergency.

It is not only appropriate but also required for the individual citizen to defend oneself and one's family.

4. Hone Navigation Skills

In an urban survival situation, knowing how to navigate both with and without technology is essential. Learn how to find your route using maps, compasses, and landmarks in case GPS and other conventional navigational aids don't work. Your confidence and preparedness will increase if you routinely practice these abilities.

And the most skilled navigators are always favored by the winds and waves.

5. Build a Comprehensive Stockpile

Your stockpile becomes your lifeline in the event that city services are compromised. Ensure you have clean water, non-perishable food, medical supplies, and survival gear on hand. Make sure you have enough supplies to last your family a long time. To keep these supplies ready, rotate and refill them frequently.

Natural disasters cannot be prevented, but we can arm ourselves with knowledge; if disaster preparedness was adequate, a great number of lives could be saved.

6. Secure Shelter Options

Decide on safe havens in your city, including safe houses or designated emergency shelters. During a crisis, having a designated safe haven to retreat to can offer a crucial sense of security. Make sure you have these places thoroughly recorded and that you are aware of the ways to them.

Have and not need is preferable than need and not have.

7. Craft an Unyielding Emergency Plan

Expert military planners and preppers stress the importance of having a well-thought-out disaster plan. Your entire life should be covered by this strategy, from making sure your house is secure to building a network of reliable contacts.

Knowing where to evacuate and doing emergency drills with your family will bring you a sense of preparedness that will come in very handy in the event of a disaster. Recall that your plan is your road map to security.

Learning via disaster is more expensive than preparing via education.

8. Embrace the Art of Urban Camouflage

It can be a tactical advantage to blend in with your urban surroundings during a crisis. Choose modest apparel instead of gaudy items that could attract unwelcome attention. Remain situationally aware and cultivate a low-key presence to avoid drawing attention.

You are planning to fail by not making any preparations.

At times of urban crisis, mental preparation is just as important as any tools or instruction. Keep in mind that a combination of preparedness, resourcefulness, and flexibility is necessary for survival. Remain alert, educated, and responsive as needed. It's essential for both your safety and the safety of those you love.

Remember these tactics as you venture out into the unpredictability of the urban landscape. Get ready now so you may confidently take on the challenges of tomorrow. Whether you're an experienced prepper or a novice to the realm of urban survival, be aware that the abilities and information you acquire now could come in very handy in the event of an emergency.

Remain alert, keep safe, and be prepared for anything.

BOOK 20: SHELTER AND SANITATION SOLUTIONS

SHELTER

Preppers have a plethora of options to purchase, construct, or create the ideal shelter to withstand any type of disaster, depending on whether their plans involve bugging out or staying put.

For everyone, staying put is the most practical course of action. This typically entails staying at home, which helps to uphold the morale of the family, with a complete supply of necessities in a cozy and familiar location.

However, a lot of individuals look for other emergency shelters with the promise of better security, seclusion, or just a place to hide until the worst passes. The only things limiting those preppers' options are their imaginations and their pockets, but we can talk about that at a later time.

Emergency Shelters

Experts have developed far more advanced survival solutions than basic backyard shacks and tents for people looking for the best. The shelters of today provide all the necessities for a week or a year of survival. Solar-powered appliances, indoor plumbing, and electricity are all included in these shelters.

DIY Shelters

For emergency shelter planning, a lot of preppers think that doing it oneself is more beneficial. To make sure your bunker is safe, there are a few things you need to do.

Make sure the underground shelter is deep enough to shield you from any potential threats before entering. Make sure you locate your shelter away from regions such as New Orleans, where high water tables render underground shelters impractical, and away from flood plains. Ensure that your bunker stays clear of any existing electrical and fuel connections, as these could endanger both your safety and the structure. Make sure there is enough ventilation, and plan for multiple exits in case one gets obstructed for whatever reason.

Compared to building underground, building above ground poses fewer issues and limitations. In terms of layout, design, and building materials, you have more options.

Bug-Out Campers and Trailers

Campers or trailers are the next best thing to sheltering in place if bugging out is your plan of action. These campers and trailers, like RVs, provide a way of escape as well as shelter. These can be turned from a standard travel trailer or camper to a self-contained survival RV by using it as a basis.

Another is the "stealth trailer" strategy, in which a prep starts with a freight trailer and customizes the interior to suit their requirements. Although it appears to be a simple box trailer from the outside, the interior is furnished with everything the family might possibly need for survival.

Think about an RV's longevity and simplicity of modification while choosing one. You are going to need to make a few changes. To stop rust, start by treating the chassis and undercarriage. Sealing off the roof will stop leaks and improve the resilience of your shelter against damp and heavy rain. To aid in keeping the weather out, insulate your shelter. The camper alone will set you back between $2,500 and $7,000 for this basic do-it-yourself shelter, plus an additional $500 for upgrades.

Cargo Van Campers

The VW bus, a classic camper's favorite, serves as the model for the cargo van conversion option. Find an ordinary cargo van to start with, and with a few adjustments, you'll have a completely mobile shelter on wheels.

The vehicle appears to be your typical work truck from the outside, but inside it is equipped with a kitchenette, bunk beds, a portable toilet, and a fridge to store food. It is simple to store food, supplies, and household items when seats are designed to fold down. Your family can stay in one of these roomy campers and be secure for several weeks until you get to safety.

Bug Out Tents

Bug-out tents come in a variety of designs, sizes, and materials, so the one you choose will depend on how you plan to travel in the event of a disaster that forces you to leave your house. Will you be loading up a vehicle with supplies, or will you be trekking on two legs and carrying a pack?

range of bugging-out from aThe Bug-Out Shelter is a basic one-person bivouac tent with a tarp draped over a rope that is secured to two trees and staked at the corner. spacious dome tents that can accommodate two to four people. These days, the majority of pack tents use lightweight poles and materials durable enough to withstand mild weather.

Larger, more robust tents with a respectable square footage are possible when bugging out with a car. These are the hunting tents made of treated canvas (or other sturdy material) that include individual rooms that can hold a wood burning stove and accommodate up to ten people for sleeping. These kinds of tents are also made with features that are intended for prolonged use by several people when bugging out.

These tents give an additional layer of weather protection in the event that you become stranded in inclement weather thanks to the detachable inner-nest feature. Their robust frameworks enable rainy setups without compromising the structure's integrity or allowing leaks. These durable tents are constructed with two layers of thick fabric and reinforced with Cordura. They can expand to 86 square feet of living area because to their redesigned feature, which can hold more people and supplies.

Long-Term Shelter

You would be better off building a cabin in the woods, preferably in a temperate area where you can raise a garden, hunt, and fish, if you're searching for long-term safety. Your ultimate objective should be to live in a spot like this—it should be remote, but not so remote that you can't still afford to rely on nature for comfort. Though there are many hazards that could make you go into prepper mode, it's crucial to keep in mind that nature has a far higher chance of recovering than any man-made structure.

Needs vs. Wants

Consider what you would possibly need to survive for a minute if you want to identify the greatest shelter for preppers. You can probably establish a slightly more pleasant existence if you strike a balance between your needs for food, water, and shelter and any concerns about protection from attack. Being ready for tomorrow entails more than just keeping a healthy sense of paranoia or planning for the near future; it also entails giving your future self some serious thought.

HYGIENE

Maintaining good hygiene is crucial for mental well-being, comfort, and a sense of normalcy. It's difficult to stay upbeat when you're itchy, odorous, and unclean. The body's capacity to ward off sickness is also diminished during periods of excessive stress. Maintaining your health now more than ever requires practicing good hygiene. One of the simplest ways to stop the transmission of illness is to wash your hands with soap and water. Although cleansing your hands with soap and water is the ideal method, using an alcohol-based hand sanitizer that contains 60% or more alcohol works just as well in their place.

EFFECTIVE HAND WASHING

- Damp hands under flowing, clean water. When available, use warm water.
- Make a lather with liquid or bar soap.
- Use your fingers to scrub all surfaces for at least 20 seconds while slowly singing the "ABC song."
- Thoroughly rinse with fresh water.

After drying with a paper towel, switch off the spigot or faucet using the towel. Place the spent paper towel in the garbage.

Wash your hands:

• Following using the restroom or assisting someone else in using the restroom. • Prior to eating or preparing meals.

• Following a nose-blow, cough, or sneeze.

• Following an animal's handling.

• Both before and after tending to a sick or injured person.

• Following the handling of waste, rubbish, or garbage.

following exposure to contaminated water, such as floodwater.

Long showers and luxurious baths may not be necessities in an emergency, as you would be wasting clean water. While maintaining your hygiene is important, taking a bath only requires a small amount of water. Your entire body can be cleaned using bath or baby wipes. Beginning at the top and moving downward, begin by cleaning the sections that are the cleanest. Your feet should come last, followed by your groin area. This is sufficient for a few days to cleanse your body of foreign bacteria, perspiration, and grime that may cause pain; it is neither as good nor as thorough as a bath or shower. Wipes aren't nearly as effective as a sponge bath, if you have extra water. In a pail of warm water, submerge a sponge or washcloth. moist your whole body with the moist sponge. Once more, start at your head and work your way down to the dirtiest areas of your body while using soap on the sponge to scrub your entire body. After cleaning the sponge, use it to rinse off your body.

Rinsing can also be done in a camp shower, albeit it could require more water. In order to warm the water for use when needed, some camp showers keep the water in black bags that may be placed outside in the sun. Additionally, buckets and two-liter pop bottles can be painted black, filled with water, and left in the sun for a few hours to warm the water. These can then be conveniently transported inside for washing, cleaning, and other purposes.

Remember to wash your teeth every day, even in an emergency. It's not simply for foul breath. Cleaning your mouth not only makes you feel better, but can you image having a toothache without access to a dentist if the problem persists for some time? Use only purified water to clean your teeth. Keep your teeth healthy and instill in your kids the value of good dental hygiene at all times.

You can feel and look better just by doing something as basic as brushing your hair. When things are stressful, remember to take care of yourself.

It could be difficult to obtain safe drinking water in an emergency. Use the water you have wisely. Make sensible use of it, but don't sacrifice hygienic practices and sanitation.

SANITATION

Clearing our bodies of waste is a crucial aspect of maintaining daily health. We must create a secure and hygienic approach for ourselves and our families to take care of this necessity in a reasonably easy and comfortable manner in the event of an emergency without access to regular facilities. When something is uncomfortable to use, many people "hold it" and wait for something better. If nothing else, they may become "stopped up," leading to serious digestive issues for which they might not have access to modern care in an emergency. Thus, prepare ahead of time and gather supplies for home sanitation in case of an emergency. However, consider these points before making your emergency potty:

Privacy: Most people have "stage fright" when it comes to using the restroom in public, which prevents them from going when they need to if they're uncomfortable. Use the current restroom, if it's available, or designate a separate room as the alternate restroom. Other possibilities are to build an outhouse, use a commercially produced privacy shelter, or block off a portion of a room or the yard with tarps or blankets to create a privacy shelter. Most people lack the space, time, money, expertise, or skills necessary to construct and maintain a suitable outhouse.

Distance: Your improvised bathroom should be both close enough to your living space for family members to find their way there and back without difficulty and far enough away to ensure privacy, safety, and a clean living environment without causing unnecessary stress. For this reason, the home bathroom can serve as an emergency potty room if it is accessible.

Usability: While maintaining appropriate cleanliness is in everyone's best interest, keep things simple. Verify that it fits your family's demands and available resources. Consider the likely users' age and skill level. Elderly family members will require something to support their body weight because they might not be able to "squat." Facilities that are suitable for their abilities or at the very least assistance in using the facilities are required for children and individuals with access and function issues.

Control of sanitation: It is not enough to simply "go behind the tree in the backyard." NEVER let trash accumulate on the ground. It draws in vermin and mice, who will subsequently spread those pathogens and illnesses throughout your living space. Keep in mind that maintaining a clean living space and yourself is the first rule. Stock up on toilet paper, feminine hygiene products, and disposable or cloth diapers for at least three months. Set up a hand washing station with a water dispenser, paper towels, soap, and hand sanitizer next to your toilet. Do not forget to routinely sanitize the dispenser and other locations that are touched a lot.

Eliminating waste or storing it temporarily: When cleaning the area and moving the bags to be disposed of, put on disposable gloves. Any garment or object that comes into contact with the excrement should be disposed of or cleaned. If you are able, dispose of it by burial; if not, store it temporarily until another method is found. NEVER dispose of rubbish near canals, streams, or other places where water may flow.

POTTY OPTIONS

By definition, an emergency is something out of the ordinary. If you didn't get a two-bucket sanitation system or you don't have your emergency supplies at home, you can find yourself without them. You can utilize safe sanitation practices in conjunction with other solutions to adjust to the circumstances you find yourself in.

"Chamber Pot": Set aside a tiny pot exclusively for use as a chamber pot. You might use a tiny bag to line it. These are more user-friendly for young children than a taller bucket and can be used both indoors and outdoors. Even if they are already trained to use the toilet, smaller children benefit more from potty training seats than from taller buckets.

Commercial Sanitation Bags: Prefilled with chemicals or absorbent materials to break down human waste, many commercial sanitation bags are available. They fit easily into vehicles and evacuation packages. Vomit can also be disposed of in certain sanitation bags, which are specifically meant for liquid waste like urine. By absorbing the liquid, these cause it to gel. Additional sanitation bags are designed specifically for solid waste and include a chemical that decomposes and renders the trash harmless. Commercial sanitation bags can be disposed of in any trash can after usage if they are utilized correctly.

Trash Can: Only use this if there is no other option. This might be utilized temporarily in a school or in an emergency situation at an office. Choose two cans, a #1 and a #2. Line the #2 can with two garbage bags, then cover the bottom with shredded paper before using.

The ideal choice is to bury the garbage, provided that you have a safe place that is at least 200 feet away from any living area or water source. Keep it out of your garden. Include digging tools in your sanitation supplies, such as shovels, picks, digging bars, and post hole diggers. Consider the type of soil you have and the possible digging season. With a shovel in hand, dig a hole that is two to three feet deep, then backfill the waste with dirt.

Book 21: Prepper's Gardening Guide

Nearly all gardening guides teach you how to garden in a "normal" setting—that is, not in an emergency situation or in an austere environment. This indicates that they are presuming that you will have simple access to clean water, fertilizer, pesticides, electricity to run grow lights, and gasoline for machinery.

Of course, you can benefit from those contemporary amenities. However, it's likely a sign that things are bad in the world and you can't rely on those comforts if you're in a circumstance where you have to rely on your garden to survive.

Crucial considerations are:

- Be able to locate, capture, store, and reuse water.

Possess the following essential unpowered hand tools:

- a shovel, hoe, rake, and wheelbarrow.

In order to have a method of converting haphazard organic waste into plant food without having to purchase it from the store, composting is even more crucial than usual.

Consider your garden from a long-term perspective, paying special attention to the soil. If you improve the quality of your soil during good times, it will continue to improve during hard times when store-bought nutrients are unavailable.

Rather than constantly purchasing new seeds at the start of the growing season, have some on hand. As the Covid-19 pandemic of 2020–21 demonstrated, seeds can be scarce during a crisis.

Undoubtedly, we've all realized that disruptions in the consumer products industry don't always require us to live in a post-apocalyptic, zombie-infested world. A little virus was all that was required. The COVID-19 pandemic has increased awareness of the benefits of cultivating a self-sufficient garden due to its food shortages and shelter-in-place instructions. However, what exactly is self-sufficiency in gardening, and how can one create a self-sufficient garden?

The Self-Sustaining Food Garden

To put it simply, your family's demands for produce can be met entirely or in large part by your own self-sufficient garden. Growing a self-sufficient garden not only helps us become less reliant on the commercial food chain, but it also feels really good to know that we can support our families and ourselves in a crisis.

Whether you've been gardening for years or are just starting out, these pointers will be useful for designing an independent garden.

The majority of vegetable plants need six or more hours of direct sunlight every day, so pick a sunny spot.

Start slowly: Concentrate on a few of your favorite crops when you first establish a self-sufficient food garden. A great first-year aim is to grow as much lettuce or potatoes as your family will consume in a year.

Maximize the growth season by planting vegetables that grow in both the warm and chilly seasons to extend the harvest season. You may grow three seasons' worth of fresh food in your self-sufficient garden by planting peas, tomatoes, and Swiss chard.

Go organic: To lessen your dependency on artificial fertilizer, compost grass, leaves, and kitchen trash. Gather rainwater to utilize for gardening.

Food preservation - Boost your garden's self-sufficiency by stockpiling the maximum amount of crops for the off-season. Grow easy-to-store items such as winter squash, onions, and potatoes, then freeze, can, or dehydrate excess garden products.

Sowing successively: Avoid planting your corn, radishes, and kale all as once. Rather, seed a modest amount of these vegetables every two weeks to extend the harvest season. This gives these crops—which could cause a feast or famine—a few weeks or months to mature.

Plant heirloom varieties: Heirloom seeds grow true to type, unlike contemporary hybrids. Using the vegetable seeds you have gathered is another step toward self-sufficiency in gardening.

Make it yourself Reusing plastic containers and making your own homemade insecticidal soaps are cost-effective ways to lessen your need on store-bought goods.

Maintain documentation Monitor your development and make use of these notes to increase your gardening performance in the coming years.

Be patient: It takes time to become completely self-sufficient in your gardening, whether you're building raised garden beds or modifying the local soil.

PLANNING A SELF-SUFFICIENT GARDEN

Are you unsure about what to plant in your food garden to sustain yourself? Try these heritage vegetable cultivars:

- Asparagus – 'Mary Washington'
- Beets – 'Detroit Dark Red'
- Bell Pepper – 'California Wonder'
- Cabbage – 'Copenhagen Market'
- Carrots – 'Nantes Half Long'
- Cherry tomatoes – 'Black Cherry'
- Corn – 'Golden Bantam'
- Green beans – 'Blue Lake' pole bean
- Kale - 'Lacinato'
- Lettuce – 'Buttercrunch'
- Onion – 'Red Wethersfield'
- Parsnips – 'Hollow Crown'
- Paste tomato – 'Amish Paste'
- Peas – 'Green Arrow'
- Potatoes – 'Vermont Champion'
- Pumpkin – 'Connecticut Field'
- Radish – 'Cherry Belle'
- Shelling beans – 'Jacob's Cattle'
- Swiss chard – 'Fordhook Giant'
- Winter squash – 'Waltham butternut'
- Zucchini – 'Black Beauty'

SETTING MENTAL EXPECTATIONS

Many novice gardeners give up because they are overly overwhelmed or disappointed, blaming the gods for their poor soil or their lack of a green thumb. However, nobody has a natural green thumb save for the Hulk, Grinch, Martian Manhunter, and Swamp Thing.

The most crucial factor is this: You have succeeded if you have managed to get plants to produce some food. You are succeeding at gardening whenever you are growing your own food, even if you could have done better.

SOME TIPS

Plan and grow your own "seed" crops

When considering seed saving, consider unconventional ideas. Many crops you can grow enough of to eat and store some to transplant next season if you save your bulb harvests or discover how they multiply.

For instance, store your best garlic cloves for planting the next growing season. You can save some of the spring potatoes to plant outside in the fall, assuming you have two planting windows for potatoes. Store a few sweet potatoes and cultivate slips for planting the following spring.

Discover how to begin from seeds, both indoors and outside.

It's a useful ability to know how to successfully produce crops from seeds. If growing inside, it might need some experience and the appropriate tools. But the amount of time and money you save by not having to buy transplants from the garden center will more than make up for the initial investment. Even though costs have increased, a packet of seeds often costs a few dollars (or is free if you preserve your own), and there are typically dozens of seeds inside. Proper storage of seeds can extend their shelf life to several years and provide more transplants than you may require.

However, a package of seeds is typically far less expensive than a six-pack of vegetable transplants (whose cost has also increased!).

It's a useful ability to know how to successfully produce crops from seeds. If growing inside, it might need some experience and the appropriate tools. But the amount of time and money you save by not having to buy transplants from the garden center will more than make up for the initial investment. Gaining knowledge about starting from seed has the extra advantage of making a large variety of fruits, vegetables, flowers, and herbs available for you to include in your garden.

Find out which perennial plants thrive in the region and in your zone.

A non-woody plant that endures multiple growth seasons is called a perennial. For instance, a plant's top may wither away during the winter and reappear each spring from its preexisting roots. The plant could also retain its leaves all year round.

With only a little seasonal care or upkeep from you, perennial crops grow back year after year. Consider including perennial plants in your garden, such as asparagus, strawberries, rhubarb, Jerusalem artichokes, perennial kale, I'itoi onions, longevity spinach, and peppers.

Plant fruit trees that are appropriate for your environment.

Find out which fruit tree varieties thrive in your area and incorporate them into your landscape. Choose fruit trees that need that amount of chill hours (or less) after finding out how many are received in your location. For instance, certain regions have native soil that supports the growth of citrus trees, as well as fig, peach, and pomegranate species.

In addition to bearing a lot of fruit, fruit trees often have deciduous leaves that contribute greatly to the compost pile and soil.

Remember to include other fruits as well, such as goji berries, grapes, raspberries, and blackberries. Find out what grows well where you live, then plant it!

Discover how to multiply plants.

There is an alternative method of adding perennials, but it can involve a significant financial outlay. Discover how to grow and propagate your current plants, as well as the plants of others. There are numerous approaches:

There is an alternative method of adding perennials, but it can involve a significant financial outlay. Discover how to grow and propagate your current plants, as well as the plants of others. There are numerous approaches:

- Cuttings
- Division
- Air or ground layering
- Grafting

Learn how to compost

Compost is a soil-improvement and fertilizer mixture made from materials that would otherwise end up in a landfill. The best approach to get your soil ready for planting in the spring, summer, and fall is to add compost to your garden at the start of each season.

Compost:

- Improves the texture of the soil by adding organic matter.
- There are billions of living microbes in fresh compost.
- Enhances the health and yield of plants.
- Guards against certain illnesses for plants.
- Moderates the pH of the soil.
- supports the soil's natural beneficial bacteria.

Learning how to compost should be at the top of your list of things to do to become more self-sufficient in the garden, as there are many reasons to add compost.

Create your own fertilizer for self-sufficient gardening

Compost is a vital soil conditioner, but occasionally you may want to supplement your soil with more fertilizer. There are two methods by which you can make some of your own:

• The high-nitrogen manure that chickens and other livestock produce can be added to compost and then fed to your plants.

• Let comfrey grow. Nitrogen, phosphorus, potassium, and several trace elements are abundant in comfrey leaves. Comfrey leaves break down quickly into a liquid, which is great for use as a liquid fertilizer.

• Plant cover crops. Certain cover crops can be used as green manure or to assist raise the amount of nitrogen in the soil.

• Grow Heat-Resistant Cover Crops Rather Take some time off this summer.

Learn how to collect and use rainwater

Rainwater collection and storage for gardening purposes makes logical and was a custom among our ancestors.

Preserving rainwater has several advantages.

- Water conservation.
- Access to water during a drought.
- The concentrations of dissolved minerals such as calcium and magnesium in rainwater are lower.
- Rainwater isn't treated with minerals, salts, or chlorine like tap water is.

Maximize your garden space for self-sufficient gardening

No matter how big our gardens are, there never seems to be enough space for us to grow everything we desire. You'll become more self-sufficient in the garden by making effective use of your available space.

Sustainable gardening and living offer an exciting route to environmental responsibility, self-sufficiency, and a stronger bond with the natural world. People can cultivate a healthy relationship with the environment and lessen their ecological imprint by adopting eco-friendly farming techniques and homesteading and off-grid living ideas.

Every action made toward sustainability—from planting native plants and creating wildlife-friendly habitats to composting and collecting rainwater—contributes to a healthier earth.

BOOK 22:
BONUS CONTENT 1: ADVANCED SURVIVAL TECHNIQUES

We have great ability and will go far. The more skills we have, the less reliant we are on technology. Although everyone appreciates having equipment and thinks it's extremely handy, abilities are lifetime if they are regularly practiced.

Most skills are things you can practice in your own backyard, at home, or at a local park. While you may need to practice in a specific area for certain skills, such as hunting or fishing, most activities may be done at home or at a nearby park.

Furthermore, try not to stress about failing. As with most things, you won't be perfect right off the get. Take your time and practice often. Think of it as a journey where every try gets you better.

There will never be a perfect job for anyone since there are too many variables in life, including the way we think.

As a result, the more we practice, the better off and stronger our foundation will be.

Practice now so that you can use the talent when you need it most, as in a crisis.

HOW TO MAKE AND SPEAR FISH

As long as you get it right the first time, crafting a survival fish spear is not that difficult. That also applies to all of your other abilities. If you do things correctly the first time, you won't have to waste time doing it again.

Here's the initial step: Trim a straight sapling to a length of eight feet and a thickness comparable to a broom handle. After that, cut the sapling in half lengthwise, keeping the prongs open with a wedge-shaped piece of wood slipped into the split. Use cordage to lash the wedge in place so that the sapling doesn't split. When using rawhide or sinew, make sure to tightly wrap it around the object to prevent any relaxation or complete loosening. The size of fish you pursue should dictate the length of the prongs. Prongs measuring six inches work well for medium-sized fish. You will then have a survival fish spear if you cut the prongs down to sharp, tapered points and fire-harden the tips.

To utilize the fish spear, raise it above your head by grabbing its upper end with your left hand if you're left-handed and your right hand if you're right-handed. The shaft is held loosely in the left hand. The right hand is utilized for all thrusts, and it serves merely as a guide. To reduce refraction, the tip is gradually lowered into the water and repositioned until it is only a few inches from the fish's back. To ensure that the next prong hits the fish if the first one misses, keep the prongs perpendicular to the fish. Similar to the heron, the last thrust consists of a single, fast jab with the right hand that pins the fish to the bottom and keeps it there.

How to Find Water in the Wild

The most vital resource you will have in any survival situation is water. You can survive for a day without eating and, unless it's extremely cold outside, you usually don't require shelter immediately away. Even though you may survive without water for a full day, going without water weakens your body and mind, making it more challenging to complete the duties required to survive. Additionally, your body will shut down and you won't be able to function for merely three days without any water.

Your body can effectively process food, circulate blood, maintain a healthy body temperature (preventing hypo- and hyperthermia), think clearly, and perform a variety of other internal functions with roughly two liters every day.

Water's vital importance to your survival is evident. Fortunately, water can be discovered pretty easily in almost any habitat on Earth with a little bit of knowledge. This section will guide you through many approaches to locate water that are suitable for tropical, frigid, and desert places, as well as varied approaches that will work for moderate climes and other environments.
While many of the following suggestions also apply to other climates, they are particularly effective in temperate and tropical regions.

Start With the Obvious: Streams, Rivers, Lakes

In the wild, these are the most visible places for you to get water. Your best bet is to use clear, running water since the movement prevents bacteria from growing. This implies that you should start your search for minor streams. Although rivers are fine, larger ones can contain a lot of pollution coming from upstream sources. Ponds and lakes are fine, but because they are stagnant, bacteria are more likely to grow there.

So, how exactly do you locate these bodies of water? Utilize your senses first. Even at a considerable distance, you might be able to hear running water if you remain motionless and pay close attention.

The next thing you'll do is look for animal traces with your eyes; they could point to water. Although irritating, swarms of insects indicate the presence of water nearby. And you might find your much-needed H2O by tracking the flight path of birds, especially in the morning and evening. It's very crucial to observe animal behavior in the desert. In the sand, animal tracks will be easier to identify, and they will nearly always lead to water in the end. In arid regions, birds will also go toward bodies of water.

Just be aware of your surroundings. Water flows downhill, thus pay attention to gullies, ditches, valleys, etc. When you get to low ground, you'll probably come upon some water.

COLLECT RAINWATER

One of the healthiest ways to stay hydrated without running the danger of contracting a bacterial infection is to gather and consume rainwater. This is particularly true in untamed, rural areas (rain initially passes through emissions, pollution, etc. in urban locations).

There are two main ways to gather precipitation runoff. Using any and all containers you may have on you comes first. The second method involves tying a tarp or poncho's corners around trees a few feet above the ground, making a depression in the middle with a small rock, and allowing the water to gather.

By tying the poncho or tarp to funnel into your bottle, pot, or whatever you have, you may combine these techniques and increase the effectiveness of your containers (as long as it doesn't overflow and waste water!).

COLLECT HEAVY MORNING DEW

Are you trying to find a method to gather one liter of water in an hour? Take a walk through meadows, long grass, etc. before daylight, and tie some absorbent garments or cloths around your ankles. When the cloths are completely saturated, wring out the water and continue. Just make sure that no toxic plants are producing dew for you to gather.

FRUITS/VEGETATION

A lot of water is found in fruits, vegetables, cactus, fleshy/pulpy plants, and even roots. For all of these, you just need to gather the plants, put them in a container, and use a rock to pound the plants into a pulp so that the liquid can be collected. Even though it won't be much, every little bit helps when things are bad.

This technique is particularly useful in tropical climates with plenty of plants and fruits. Coconuts are a great way to stay hydrated. However, unripe, green coconuts are actually preferable because the laxative properties of ripe coconut milk would cause you to become even more dehydrated.

COLLECT PLANT TRANSPIRATION

Using plant transpiration as a source of water is another simple solution. This is how moisture moves from the roots of a plant to the underside of its leaves. After that, it evaporates into the atmosphere, but not before you collect the water.

Tie a bag (or something you can make into a bag; the bigger the better) around a bush or branch of a lush green tree first thing in the morning. To provide the water with a place to collect, add a pebble inside the bag to slightly weigh it down. The plant transpires and creates moisture throughout the day. But instead of dissipating into space, it gathers at the bottom of your backpack. When dealing with toxic plants, never do this.

TREE CROTCHES/ROCK CREVICES

Similar to fruits and vegetables, this source also won't supply a lot of water, but it's still useful in desperate situations, especially if you find yourself alone in a desert. Small areas where water can accumulate are the crotches on tree branches or the cracks in rocks. Even when water isn't visible, bird droppings near a rock fissure in an arid region may be a sign that water is there. A piece of cloth or clothing should be inserted into crotches and fissures to absorb any moisture and then wrung off. Try again, and come back for more after a downpour.

DIG AN UNDERGROUND STILL

Building a still has the advantage of giving you a consistent, sizable supply of water (in comparison to other techniques), and you can better plan and ration as you will have a better idea of how much you will be receiving.

Sills come in aboveground and subterranean kinds. If you're running low on energy and can't dig a big hole, the aboveground variety might still be helpful. However, the underground variety is your best option because it collects more water.

Directions for your underground still:

Supplies

- Container (the largest you have)
- Clear plastic sheeting
- Digging tool
- Rocks
- Optional: something to act as a drinking tube/straw (CamelBak straw, bamboo/other plant)

Instructions

Choose a spot where the majority of the day is sunny.

Dig a pit that is 2 feet deep and 3 feet wide. Within that, dig a second, smaller hole for the container.

Not required: Fasten the drinking tube to the container's bottom. Proceed to the next step without one.

After setting the container within the pit, raise the tubing and exit the hole.

To seal the plastic covering over the hole, fill it with soil and rocks.

Place a little rock in the middle of the sheet so that it hangs over the container and forms an inverted cone.

Use the tube directly if you have one. If not, retrieve the container from the bottom and replace it when the water has been placed somewhere else.

At that depth, moisture is nearly always present in the ground. This will cause condensation in response to the sun's heat, which will gather on the plastic. That condensation is forced into your container by the inverted cone. You should anticipate a crowd.5 to 1 liter every day, therefore multiples would be required to cover a full day's worth.

Melt Snow and Ice

Snow and ice are plentiful far into the summer, and occasionally throughout the year, especially in the highlands. If you're near or on the ocean in a polar zone, seek for "ancient ice," or ice that has been through showers and thaws, and icebergs for a source of fresh water. Freshwater ice splinters easily with a knife and has a bluish tint and crystalline structure, whereas salty ice is opaque and gray. If you find yourself in a boat with salt water all around you, collect some of it in a container and let it freeze. The salt will gather as slush in the center, whereas the fresh water will freeze first. Take out the ice and throw away the goo.

Although snow and ice make great sources of water, they should always be thawed and cleaned before use. Your body temperature will drop if you eat only snow or ice, which causes your metabolic rate to increase in an attempt to stay warm and causes dehydration.

The easiest method to melt snow or ice and make it taste good is to just combine it with some additional water, even a little bit of it, and swish it about until the snow melts. In order to avoid scorching the snow or ice and creating an unpleasant-tasting drink, dilute it with a small amount of other water if you are heating it.

Avoid Water Substitutes

You might be tempted to attempt non-water liquids in a dire survival situation instead of the genuine thing. These are to be avoided in all except the most extreme cases. Non-water replacements, in general, simply make you feel less energetic and healthy. These alternatives, together with the detrimental traits they have below:

- Alcohol. Dehydrates and clouds judgment.
- Urine. Contains harmful body waste and is about 2% salt.
- Blood. May transmit disease. Also has high salt content.
- Seawater/Sea Ice. Contains 4% salt. It takes more water to rid your body of the waste from seawater than what you get from it. Just depletes your body's H2O supply.

If you're a fan of Bear Grylls, you are undoubtedly aware of his well-known urine-drinking incident in the Sahara Desert. Is it safe to do this, or was that just for fun? Urine is a last-resort survival tactic that can let you survive for a few more days. It is 95% water, with the remaining 5% being waste products that, if consumed for an extended length of time, will eventually cause renal failure. Naturally, this technique gets riskier the more dehydrated you are.

Try all of the aforementioned options before drinking your own urine. You have a high chance of finding real H2O if you put in a little work, research, and creativity.

PRACTICE KNIFE THROWING

Your fingertips slide easily over the cold steel as you fix your concentrated gaze on the target, which is around 13 feet away. You focus all of your concentration on the 12-inch throwing knife, and it shoots toward the bullseye, which is about 4-inches wide. The world disappears.

Throwing a knife at a target takes talent, practice, and attention that few things in life can match in the 2-second sequence. Although it may seem challenging at first to start throwing knives for fun or competition, it's not that hard to get started.

This is all the information you require to begin throwing knives.

Reasons to Take Knife Throwing Classes

As you read this, you might be debating whether or not to take up knife throwing, particularly if others close to you don't share your enthusiasm. These are four strong arguments for getting a throwing knife.

1. Knife Throwing Is A Historical Practice

It's likely that if you love knives, you also love history. With just a target and some throwing knives, you may relive the fascinating history of the art of knife throwing, which dates back to the prehistoric era. Native Americans, African tribes, and Japanese warriors all employed throwing knives and other tools.

2. It Is Competitor-Friendly

Knife throwing is a sport as well, but if you're simply down to hang out in your backyard and let off some steam, go ahead. The top knife throwers compete in regional tournaments held across the nation by numerous organizations, including the American Knife Throwing Alliance. Knife throwing can provide you with a little competition, which can serve as motivation to be the greatest.

Third, Knife Throwing for Fun

Knife throwing doesn't have to be intense, even if this one might seem obvious. Like playing pool or shooting darts, it may be an enjoyable and soothing pastime. A decent set of throwing knives and a target can provide hours and hours of fun instead of wasting money on going out or finding transient amusement.

4. It Is Often Quite Social

Knife throwing is a terrific way to be social and meet new people, but it can also be a great way to relax and spend time alone. You can host casual get-togethers in your backyard or participate in tournaments. You may easily bring your buddies over and teach them how to throw knives if it doesn't sit well with you.

How Long Does it Take to Learn Knife Throwing?

This one is contingent upon your level of consistency and weekly effort. However, if you train six hours a week, you can usually learn how to throw a knife in six months with continuous practice.

Basic Types of Throwing Knives

It's vital to examine the various kinds of throwing knives before proceeding to the instructional portion of this article. Other throwing weapons exist, such as the tomahawk and shuriken, but as you go, you can learn more about them. The three main categories of throwing knives are listed below.

1. Blade-Heavy Throwing Knife

Each type's name is rather self-explanatory. A throwing knife that weighs more in the blade than the handle is said to be blade-heavy. You would grasp the handle of this kind of knife when throwing, even if it can be a little perplexing at first, because you want the heaviest end to be thrown first. Because they're simpler to learn how to throw using the hammer technique, blade-heavy weapons are sometimes seen to be the best option for novices.

2. Heavy-Handled Throwing Knife

The majority of a handle-heavy throwing knife's weight is located in the handle, hence you should throw the knife with the handle going first. It could be difficult for beginners to get used to this because you have to grasp the knife's blade end.

3. Balanced Throwing Knife

While the center of gravity of the two previously mentioned types of knives is not in the middle, balanced knives have it. This allows you to throw without fear of injury from the handle or the blade. Although balanced throwing knives have more consistent spins, some novices first find these models difficult to use. These nevertheless provide a great deal more versatility.

How to Throw a Knife

Now that you have a better understanding of the benefits of learning knife throwing as well as the many kinds, it's time to get started. The fundamentals of throwing a knife are as follows.

Using the Hammer Grip first

When throwing a knife toward a target, there are a few different ways to hold it, but we'll start with the most common grip: the hammer grip. For novices, the hammer grip is by far the simplest. You grip the knife in the same hand as you would a hammer, as you could imagine. For added control, wrap the fingers of your dominant hand around the knife's handle and place your thumb on top of the handle.

Put Yourself in the Correct Position
Get ten feet or so away from your target. If you are right-handed, make a 45-degree angle with your feet by placing your left foot front and your right foot back. Make sure you're stable and bend your knees just a little.

Keys to Getting the Throw Down
Once more, if you are right-handed, point your left hand as though you are aiming at the target. Raise your right hand's knife straight back above your head. Return the knife to its original position as if you were a butcher chopping meat. Leaning forward, release your grip on the knife. Knowing when to release the knife is one of the hardest tasks, but the secret is to practice and throw consistently. Adjust your release timing and throw slightly as you practice to address any problems.

Some abilities might not be applicable to you at this time, but remember the concept. You've already done your research, so you won't be as afraid to get started if the time comes for you to start keeping bees. In fact, discovering new interests could really help you discover passions you never knew you possessed.

BOOK 23:
BONUS CONTENT 2: ADVANCED SURVIVAL TECHNIQUES 2.0

How to Read a Topographic Map

All of us are acquainted with maps. These are the crisp papers that unfurl to the point where our view of the road is totally obscured when we take them out of the glove compartment. They make us wish we had two sets of eyes, one focused on the road and the other on the prearranged course, and go-go-gadget arms that stretched to their widths. They give us the impression that the GPS system's woman hadn't directed us to a dead end.

Maps have varying functions. A standard road map is the best option if you're trying to drive from Point A to Point B and would rather follow a map than GPS directions. However, you must be able to view the topography and land contours if you have deviated from the path, possibly on a hiking excursion. This implies that the ability to read a topographic map is a prerequisite.

What distinguishes a standard map from a topographic map? In summary, you may visualize a three-dimensional landscape on a two-dimensional surface by using topographic maps. These maps display the topography, vegetation, bodies of water, mountains, valleys, and other features of the area. They can be distinguished from other maps by their contour and elevation information.

The first topographic map was created by the U.S. Geological Survey (USGS) in 1879, and they are still produced by them today. These maps may now be produced considerably more accurately and quickly than they could in the past when they were made by hand because to advancements in satellite imaging and aerial photography. The USGS had to invest a lot of money and effort in mapping public land when it initially began the task. The only means to get to the mostly unpopulated West was by mule pack train, and the equipment used by cartographers were antiquated by today's standards. Using drawing boards and sighting instruments, mapmakers would scale the highest point in the area to get the best view point. After that, they would draw their maps using the observable and quantifiable features. This called for courage as much as expertise. The development of airplanes in the 1940s contributed to the advancement of mapping methods.

Topographic Map Lines, Colors and Symbols

In contrast to a standard map, which mostly displays roads and highways, a topographic map offers a more accurate representation of the terrain.

A topographic map's features include:

Roads, buildings, urban development, borders, railroads, and power transmission lines are examples of culture.

Water in marshes, rivers, lakes, streams, and rapids

Relief: depressions, slopes, valleys, and mountains

Vegetation: vineyards, orchards, cleared and forested regions

Toponymy includes the names of places, water features, and highways.

The maps have many applications because they display a great deal of information. Topographic maps are used in engineering, conservation, environmental management, urban planning, public works design, and outdoor pursuits including hiking, camping, and fishing.

Understanding how to comprehend the lines, colors, and symbols on a topographic map is the first step in learning how to read one. Large swathes of green indicate vegetation, blue indicate water, and gray or red indicate heavily populated areas on these maps. Tiny black squares make up houses. Large structures, like your neighborhood mall, will be visible as their true forms.

On a topographic map, lines can be dashed, straight, curved, or a combination of these. These lines show streams, highways, boundaries, and more. These lines come in a variety of hues, including brown, blue, red, black, and purple. Every hue has a distinct meaning.

Despite using symbols to making the map less cluttered, a topographic map is nevertheless jam-packed with information. On its topographic maps, the U.S. Geological Survey (USGS) uses the following symbols:

- Boundaries
- Buildings and related features
- Coastal features
- Contours
- Control data and monuments
- Glaciers and permanent snowfields
- Land surveys
- Marine shorelines
- Mines and caves
- Projection and grids
- Railroads and related features
- Rivers, lakes and canals
- Roads and related features
- Submerged areas and bogs
- Surface features
- Transmission lines and pipelines
- Vegetation

Topographic Map Contour Lines

The most important characteristic that sets a topographic map apart are its contour lines. Drawn on a map as lines joining places of equal elevation, contour lines indicate locations where the elevation would stay constant if you physically followed them. The terrain's shape and height are depicted by contour lines. They are helpful because they depict on the map the topography, or shape, of the land surface. Here's a neat method for learning how to read contour lines: Point a red laser pointer along the side of an object, such as a ball or a stack of laundry. The line you perceive will resemble a topographic map's contour line.

Topographic maps only display lines for specific heights in order to keep things simple. The distance between these lines is uniform. This distance is known as the contour interval. For instance, if the 10-foot contour interval is used on your map, you will see contour lines at 0, 10, 20, 30, 40, and so on, for every 10 feet (3 meters) of elevation. Depending on the terrain, different maps employ varying intervals. For instance, if the overall topography is significantly high, the map may be displayed in intervals of 80–100 feet (24–30 meters). This facilitates map reading because it would be challenging to deal with too many contour lines. To determine the contour interval of your map, look in the margin.

Every fifth contour line on topographic maps is an index contour, which makes the maps easier to read. Only the index contour lines are labeled because it is not practical to record the height of every contour line on the map. Compared to the ordinary contour lines, the index contours have a wider or darker brown line. The altitudes are only indicated on the contour lines of the index. Take note of the distance between lines to determine heights. You're dealing with a steep slope if the contours are near to one another. The landscape is comparatively level if there are large gaps between the contours or if there are none at all.

TOPOGRAPHIC MAP SCALE

Scale is an additional aspect of a topographic map that you should comprehend. Maps are obviously not life-sized. If not, our bags would never be able to accommodate them. Rather, one measurement on the map equals a larger amount in the real world when cartographers plot maps using a ratio scale.

There is always one at the top of the scale. It is your measurement unit, which is often an inch. The ground distance is the second figure. A map produced by the United States Geological Survey (USGS) with a 1:24,000 scale, for instance, indicates that one inch on the map corresponds to 24,000 inches (2,000 feet or 610 meters) in the real world. The scale legend on your map will always be at the bottom.

A common scale for USGS topographic maps is 1:24,000. Metric maps have a scale of 1:25,000, meaning that one centimeter is equivalent to 0.25 kilometers. With very few exceptions, the majority of the United States is mapped in a 1:24,000 scale. For instance, Puerto Rico's maps appear as 1:20,000 or 1:30,000 because the island was first drawn on a metric scale. Due to its size, the majority of Alaska maps at 1:63,360, but a few states map at 1:25,000. However, Alaska's more populous regions map at a standard 1:24,000 or 1:25,000 scale.

Buildings, campgrounds, ski lifts, and other features are all shown on a big, detailed 1:24,000 map of the area. On a map this size, you might also see private roads and footbridges.

At this scale, the 48 contiguous states, Hawaii, plus the territories would require roughly 57,000 maps [source: streets, power and water lines, and sewers]. These maps are commonly scaled at 1:600. A lower size map, such as one at 1:250,000 scale, is preferable if you wish to see a single huge area in less detail on one sheet.

HOW TO USE A COMPASS

One of the most important pieces of survival equipment is a compass, which most people carry into the mountains without really understanding how it operates.

Luckily, using a compass and a good map together won't be too difficult. Naturally, with a little expertise, it also becomes an effective navigational tool.

Essential Parts of the Compass

It will be difficult to discuss compass usage without first having a clear knowledge of the functions of each component.

So let's just go over that.

To assist you calculate distance, you can utilize rulers in addition to your map scale.

When taking a bearing, a direction of travel arrow serves as a helpful reminder of which way to point the compass.

The large arrow we'll use to assist you in centering the bezel is known as the orienting arrow.

A convenient declination scale seen on some compasses facilitates declination adjustment.

The rotating bezel, which is marked with degrees from N clockwise up to 360, is the large circular dial with degrees.

Where you'll read your bearings is indicated by a small marker known as the index line.

The magnetized needle inside the bezel is the first thing you'll notice; it always points in the direction of the magnetic north pole rather than the genuine north pole.

Our orienting lines are used in conjunction with the orienting arrow to assist you in aligning the compass with north on a map.

The magnetic needle is housed in a liquid capsule with a spinning housing. Accurate readings are easier to obtain since the liquid helps to moderate the movement of the needle.

There is always at least one straight edge that you may use to get your bearings later on, and the base plate is transparent so you can see the map underneath it.

SETTING DECLINATION

The angle that separates magnetic north from true north is known as declination. Your compass needle will only point in the direction of magnetic north. Currently, it is pointing towards a location in the northern hemisphere rather than Santa's workshop. Navigation can be seriously affected if your compass is not adjusted to take into account the difference between magnetic and true north.

The declination will be a specific number of degrees to the west or east, depending on where you are in the nation. In the declination diagram next to the legend, you can see the direction, a number of degrees, and the date of the most recent update to the map. If it's been a while since you updated your map, it could be worthwhile to check the declination online.

Once you know the declination for your area, you may enter it into your compass. Different compass brands set declination in different methods. The north marker on the bezel will point in the direction of true north when you accomplish this and the needle aligns with the orienting arrow.

The orienting arrow must be moved 16.01 degrees to the east due to the declination of 16.01 degrees east on the map above. You can now use your compass with your map when the declination has been set.

TAKING A BEARING FROM A MAP

To begin with, a bearing gives a direction a degree description. Using a bearing is far more accurate than simply indicating east or west. Alternatively, you may argue that in order to reach the lake, take a 79-degree bearing.

Using the bearing is most frequently used when hiking and you want to get to a specific location, such as a peak or your campground. It's easy to do if you know where you are on a map.

Locate your current location on the map first. Assuming we are at point A, align your position on the map with the straight side of the base plate using your compass. Next, identify the destination you wish to reach. Assuming it is point B, turn the compass around until a line connects point A and point B. Verify that the trip arrow is pointing in the peak's direction. After the compass is positioned, turn the bezel until the map's north and south lines coincide with them. To assist you place the lines exactly right, you can use the map's edge.

Make sure the North marker on the bezel points northwards on the map, not southwards. To read the bearing, glance at the index line once the orienting lines are going north-south. Assume it is 145.

Congratulations! You may now follow the bearing using the compass. With the travel direction arrow facing away from you, pull the compass forward in front of you. Now turn your entire body until the magnetic needle's north side is inside the orienting arrow. You can recall "red in the shed," if it helps. You are facing your destination and the bearing of 145 once you have turned red in the shed.

Taking a Bearing in the Field

You may also get a bearing off a nearby landmark if you're out climbing and want to see where you are on a map.
After that, you may use your map to transfer that bearing and determine your location.

Locate a landmark that you can recognize on your map first. Aim your compass so that the direction of travel arrow points straight at the landmark and away from you. Once the magnetized needle is inside the orienting arrow or the red is in the shed, maintain the compass flat and turn the bezel. You've grasped the bearing once the needle is inserted.

Accurate direction finding is simplified with a compass and seeing mirror. Point the compass toward the object while holding it at eye level. Once you can look straight down into the bezel, tilt the mirror. In this manner, the item and the bezel are seen simultaneously. You can transfer the bearing to the map after you have it.

Arrange your compass such that its edge aligns with a recognized landmark on the map. Verify that the journey arrow is facing the landmark in the correct direction. Rotate the entire base plate such that the orienting lines run north-south with the edge of the compass aligned with the item. Furthermore, the bezel's North marker points in the direction of North on the map. After you're done, you can draw a line parallel to your compass's edge on the map. You fall somewhere in this spectrum.

Triangulation

Triangulation is the procedure you use to determine your exact position. All you need to do is use a different item to take another bearing. That isn't even close to sixty degrees to the original. Your position is where the two lines intersect. You will get a small triangle if you choose the third bearing. Additionally, the reading is more precise the smaller the triangle.

Although compass skills can seem difficult at first, they become much simpler with practice. Try them out for additional practice on your preferred trail.

How to Raise Honeybees in Your Backyard

1. Choose the location

Bees require four items. They need sun first, or afternoon shade if it's really hot outside. Secondly, fresh water must be available to them close to the hive. Every day, we replaced the water in the big plant saucer that had stones in the middle for the bees to settle on. Additionally effective would be a shallow bubble fountain. Third, wind protection is necessary for the hive because it can force rain or snow inside, which makes it more difficult for the bees to maintain the hive's warmth. Lastly, bees require seclusion. The hives should not be placed close to pet areas, playgrounds, swimming pools, or busy streets. Each hive should have plenty of room; if there isn't enough room, place the hive so that the entrance is next to a tall fence or hedge. Fifty feet away from busy places is preferable. This will compel them to fly overhead in order to avoid coming into contact with people or animals. Additionally, keeping them out of sight will satisfy both humans and bees.

2. Prepare the location

Hives should be maintained off the ground to protect them from moisture and animals, and they should be facing south if at all possible. We leveled the ground, removed the undergrowth, and poured a cement pad to make maintenance simpler.

3. Put the hives in place

It's time to put your bees in their hives in the spring, when growing flowers provide a food source. The best course of action is to trust your source for installation instructions after you've decided how to purchase them.

4. Give the bees food.

young colonies have a lot of work ahead of them, including caring for the queen and her young brood, storing pollen and nectar, and caulking all the gaps and seams in their new home. We gave them some "nectar" to help them adjust more easily. The recipe is as follows: Fill the quart jars with a solution made of equal parts water and granulated sugar. Place the feeder lids on top and turn the jars so they are inside out. The lids should be just slightly damp, not dripping. The nectar from the lids will be consumed by the bees.

Our nucs first drank almost 3/4 of a quart jar per day. It gradually decreased over the course of the following three weeks to the point when we recognized sugar water was no longer required. Flowers were providing food for the bees. Moreover, sugar water should not be used if it is not necessary because it produces insipid honey.

5. Examine the hives from the inside out.

Beekeeping is mostly about observation and reaction. For a few months, if you are a new beekeeper, check the hive once a week or so to gain experience. Once you're at ease, change your schedule to once every two weeks. Verify that there are no ants on the hive, the landing board is clear of trash, and the outside of the hive is clean and clear of bee feces. On warm days only, open the hives and look within the frames for larvae and eggs. There should be an abundance of larvae in different phases of development if the queen is in good health.

Speak with a specialist if you fail to find any indications of a healthy queen. A useful resource is the beekeeping guild in your area.

In the end, the hive's health will benefit from fewer inspections. To keep the bees quiet, you must smoke the hives before opening and thoroughly inspecting them. The bees require around a day to recuperate from this stress. You won't need to pull many frames to understand what's going on within as you get more knowledge. And just watching the bees as they enter and exit the hive can reveal a lot to you.

6. Consistently look for illnesses and pests

The most common pest in hives is varroa mites. If left untreated, they have the potential to severely harm and ultimately destroy the hive (for guidance on identifying and managing mites, refer to Pest Control, below). The little hive beetle and the wax moth are two other pests to keep an eye out for. The diseases to watch out for are European and American foulbrood. A live hive or a dead hive can frequently be determined by early intervention.

7. When required, enlarge the hive

Begin with one body-brood box deep in the hive. Place a second brood box on top once the bees have filled it with seven or eight frames of bees and brood. Permit the bees to populate the second brood box with more brood cells. Once the second brood box is nearly full (seven or eight frames of bees), place the honey super (the box where you will gather the majority of your honey) on top of it, along with a queen excluder if you choose to use one.

Pest control

Bees resemble beautiful spun sugar balls in the air that are packed with honey. Everybody is itching to devour them. These three pests represent our worst opponents along with the strategies we employed.

Ants.

By stealing honey and devouring the young, Argentine ants can destroy a colony. Since spraying would kill the bees as well as the ants, we were unable to eradicate them. With some success, we used Terro ant bait, which are small containers filled with a pleasant material that attracts ants along with boric acid. Ultimately, we found that using a physical barrier worked best. We submerged the legs of the hive stands in water-filled plastic tubs so the ants could not pass through them.

Small hive beetles

Larvae of hive beetles consume every part of the hive, including the young bees. We have been experimenting with traps like Beetle Eater and killing the beetles on the spot.

Mites of Varroa

These mites are the most destructive pests a beekeeper has to deal with since, once established, they pose a threat to the hive's life. They drink the blood of mature bees and deposit their eggs in brood cells, where their larvae feed on the young bees, weakening and eventually killing them with viral infections.

Control methods

1. Observation

You may get a decent estimate of the infestation in a hive by counting the natural mite fall during a 24-hour period. Apply cooking spray or petroleum jelly on the bottom of your Country Rube board in order to capture mites. Then, slip it into the lower portion of the board, wait a day, and then remove it to count the mites. If there are more than 10 mites in each brood box, something is wrong.

2. Dusting with sugar

You can both manage and count the mites with the powdered sugar method. After sifting one cup of powdered sugar each brood box, brush the mixture into the hive and over the tops of the frames. The powdered sugar causes the mites to release their hold on the bees, and as a result, more mites are dislodged by the bees when they brush the sugar off their bodies. Once again, collect the fallen mites using the bottom board. After dusting, ten minutes should not reveal more than a few mites. You have an issue if there are more.

3. Using mite traps

Moreover, drone frames will aid in varroa mite capture. The purpose of these frames is to stimulate bees to produce larger drone comb cells than worker comb cells. Drone combs work well as mite traps because varroa mites prefer drone broods of one to ten. 24 days after the eggs were laid, just before the drones hatch, destroy the drone comb (you can cut it out or freeze it and return it to the hive) and replace the drone frame for the upcoming cycle. (Our queens do not require the drones to reproduce because they have already mated and have an infinite supply of sperm.)

4. The Apiguard

a gel created from the oils of thyme plants and infused with thymol. It tastes like mouthwash, but it does the trick when it comes to preserving honey.

5. Formic acid

Formic acid is more toxic than thymol and gasses the mites to death. As a result, it is applied in the fall and winter, when the flow of nectar has slowed or halted, rendering the honey inedible to humans. When applying it, a respirator is required.

Honey collection

The first summer we were fortunate enough to gather honey. The bees usually build up their hive in the first year, and if they overwinter successfully, you can start harvesting in the late spring or early summer of the following year. We had four frames filled to the brim with honey, each weighing almost eight pounds, three months after we brought our bees home. Since we didn't have a professional extractor, we employed this low-tech technique.

1. Chop and grind. We scraped the honey, wax and all, off the foundation with the bench scraper and balanced the frame on a wooden spoon that was placed like a bridge across the bowl. The honey and wax in the bowl were then crushed with a wooden spoon.

2. Pulling in and adjusting. We filled our food-grade plastic bucket with this wax and honey mixture after straining it through a stainless steel strainer and two layers of cheesecloth. After that, we gave it a few days to drain and settle (bubbles and foam rose to the surface).

3. Filling bottle. We prepared our jars and spread newspapers around the floor. The honey was then released into each jar by unscrewing the honey gate, which is the stopper at the bottom of the bucket. The lids went on, and the honey went in. That was all there was to it. We harvested 12 pounds, 10 ounces of honey from 4 whole frames of honeycomb. The remaining wax was washed and then frozen. We then utilized the wax for craft projects like lip balm and hand salve, rendering it in a solar wax melter. Later in the summer, we experienced another unexpected harvest, resulting in approximately 31 pounds of fragrant, pure honey.

BOOK 24: BONUS CONTENT 3: FAMILY PREPAREDNESS

Natural disasters frequently occur suddenly and violently. They can be traumatizing for kids and terrifying for adults. It can be necessary for your family to move out and alter your daily schedule. Prepare yourself to provide your kids advice that will make them feel less afraid.

Being prepared begins with each of us doing our part to keep safe in the lead-up to, during, and following calamities or disasters. Your family and emergency responders can both be kept safer if you take the time and make the necessary preparations at home. The Red Cross offers a wealth of materials to assist children and their adults in improving home preparation.

IDENTIFY HAZARDS

Determine which catastrophes are most likely to occur in your region and study about the best ways to be ready for each.

Find out about the warning systems and signals (sirens, text messages, etc.) in your town.

Think about getting a weather radio from the National Oceanic and Atmospheric Administration. Offical alerts, watches, advisories, predictions, and other threats are broadcast over this radio 24 hours a day, 7 days a week.

Determine whether local nonprofits, Red Cross/Red Crescent, and other groups handle emergencies, and be aware of their contact information.

Learn about any disaster plans that your children's school, place of employment, or other locations where you and your family spend time are in place.

HOLD A FAMILY MEETING

Have a family meeting to go over the reasons being organized is crucial.

Go over the most common disaster kinds and provide advice on what to do in each case.

Give each family member a task to do and make plans to collaborate as a team.

Choose the places where you will gather in the event of a disaster:

- In the event of an unexpected emergency, like a fire or earthquake, outside your house and the surrounding area.
- In case you are unable to go back home, outside of your city.

Make a family evacuation plan and talk about what to do in the event of an evacuation.

If a family member serves in the armed forces or travels regularly, make plans for how you would react if a disaster occurs while they are away.

Make a strategy for family communications and record how your family will communicate in the event of a disaster.

Make necessary adjustments to your strategy if any family members have special requirements or a handicap.

Make sure your plan takes your family's pets into account.

PREPARE

Put together a disaster supplies kit.

Determine in your house where each kind of disaster has a safe place to be.

Ascertain the most effective escape routes from your residence.

Learn CPR and first aid techniques.

Teach all members of the family where to turn off the gas, electricity, and water.

Make a thorough inventory of everything in your house.

Show your family members where to look for and how to use fire extinguishers.

List on the refrigerator the emergency contacts (friends, family, neighbors, police, fire, etc.).

PRACTICE YOUR PLAN

Every six months, or so, have family practice using your strategy.

Every three months, examine the supplies in your disaster pack.

Every three months, replace food and water that has been stored.

As things change, update any emergency contact information.

How to Guide Your Children During a Disaster

Your Response Could Influence Your Child's

Although it's normal and natural to be afraid, your kids will look to you for guidance amid a disaster: Your child may grow more afraid if you act alarmed because they will interpret your worry as evidence that there is a serious threat.

Your child might experience their losses more keenly if you appear to be overwhelmed by them.

If you can show your child that you're composed and in charge, they might feel more capable of handling situations and feel more confident.

When a Child Is Fearful, They Are Fearful

The emergency could seem more serious to your youngster than it is. Children's imaginations might amplify their anxieties, so you should be mindful of these emotions. Reassurance can come from your words and deeds, so make careful to paint an honest and accurate picture.

How to Help Your Child After a Disaster

What to Expect

Youngsters rely on well-known schedules: waking up, eating breakfast, going to school, and playing with friends. They might experience anxiety, confusion, or fear when this routine is disrupted by an emergency. These emotions can show themselves in a number of ways, such as increased shyness or aggression, or clinginess or withdrawal. It's possible for your child to revert to habits they once outgrew, like carrying a plush toy or sucking their thumb.

How to Proceed.

Asking your youngster about his or her thoughts will help you focus on meeting their emotional needs once the danger has passed. Including kids in your family's rehabilitation activities will give them hope that things will soon return to "normal."

Keep young children away from television news coverage of the incident while they heal. The pictures can be extremely distressing, especially for a small child who doesn't understand they are watching old film rather than brand-new footage.

BOOK 25:

BONUS CONTENT 4: PREPPER'S RESOURCE DIRECTORY

The top communities, message boards, directories, and discussion boards about survival, handpicked from thousands of online forums and sorted by member count, popularity, and active threads.

EXPLORE THE TOP SURVIVAL FORUMS AND DIRECTORIES TO FOLLOW

1. Subreddit - Survival

Located in San Francisco, California, the Survival subreddit delves into Wilderness Survival, encompassing philosophies, knowledge, techniques, and actions tailored for short-term survival scenarios in the wilderness. Join the conversation at www.reddit.com/r/Survival/?utm_source=feedspot.

2. Survivalist Community

Join the Survivalist Forum, a vibrant community dedicated to survivalists and enthusiasts. Engage in discussions covering topics like Farming, Gardening & Homesteading, share your DIY projects, collections, hobbies, and more at www.survivalistboards.com.

3. Prepper Enclave

Be part of the Prepper Forums community, catering to Survivalists, Doomsday Preppers, and Campers. Share your survival journey, discuss gear, food storage, and exchange prepper tips at www.prepperforums.net.

4. My Survival Hub

Welcome to My Survival Forum, the fastest-growing survivalist community. Connect with like-minded individuals, exchange information, and prepare for life without modern conveniences. Join discussions on a variety of survivalist skills, stories, and ideas at www.mysurvivalforum.com.

5. Survival Magazine Exchange

Explore discussions on survival at the Survival Magazine Forum. Engage in conversations about preparedness, preppers, survival food, kits, SHTF, homesteading, and more. Join discussions on Survival TV Shows & movies, read news & books, and talk about primitive skills at www.survivalforum.survivalmagazine.org.

6. Survival Monkey Exchange

Introduce yourself in the Survival Monkey Forums, dedicated to modern survival and preparedness. Share insights on survival medicines, discuss Survival and Primitive Technology, and exchange information on wilderness and urban prepping at www.survivalmonkey.com.

7. Survivalist UK Community

Participate in discussions on frugal living and survival at the Survival UK Forums. Delve into topics about civil unrest, natural disasters, and society breakdown. Get suggestions on survival kits and all aspects of prepping at http://forum.survivaluk.net/.

8. Doomsday Prepper Hub

Join the Doomsday Prepper Forums to discuss various man-made and natural disasters, financial collapse, and ideas on food storage, farming, and crop cultivation. Connect with the community at www.doomsdayprepperforums.com.

9. Prepared Society Exchange

The Prepared Society Forum is a dedicated space for preppers, survivalists, and enthusiasts. Introduce yourself, share goals and experiences, and engage in discussions about collections, gear, and DIY projects at www.preparedsociety.com/forums/.

10. Northeast Shooters Forum - Survival

Explore survival tips and tricks at the Northeast Shooters Forum. Share your survival journey, skills, and ask questions on different survival methods at www.northeastshooters.com.

11. Survival & Self Reliance Exchange

The Survival & Self Reliance Forum welcomes individuals of all beliefs and backgrounds to discuss survival. Stay updated on news, information, and engage in discussions about survival methods with experts at https://forum.theorganicprepper.com/.

CONCLUSION

Building resilience, adaptability, and self-reliance while skillfully navigating through difficult circumstances requires a strong survival attitude. People can dramatically improve their odds of survival in a variety of situations by focusing on developing mental toughness, positive thinking, and a proactive attitude to preparedness and training.

Technical proficiency and physical prowess are not the main aspects of survival skills. They have a strong foundation in mental capacity. Adopting a survival attitude entails accepting mental toughness, adjusting to change, and keeping an optimistic view despite hardship.

People might find their inner power and realize their full potential by adopting a survival attitude. They can learn how to make wise decisions, handle obstacles with grace, and stay composed in crisis circumstances.

Recall that developing a solid survival mindset requires constant effort. It necessitates consistent practice, introspection, and a dedication to one's own development. Therefore, cultivate a survival mindset, believe in the efficacy of positive thinking, and arm yourself with the required knowledge and preparedness. You can meet any issue head-on and improve your chances of survival if you have the correct mindset. Resilience, adaptability, and self-reliance are essential for prospering in the face of difficulty.

BONUS

BONUS 1. ADDITIONAL BOOK: **The Prepper's Mindset Psychological Strategies & Emotional Preparedness for Long-Term Survival**

BONUS 2. **10 Ways to Clean Drinking Water After Disaster**

BONUS 3. Beekeeping How To Start Beekeeping In 2024

BONUS 4. 11 Foods To STOCKPILE That NEVER Expire!

BONUS 5. **The Survival Pack | Doomsday Preppers**